MEMOIRS
of the
American Mathematical Society

Number 999

On Systems of Equations
over Free
Partially Commutative Groups

Montserrat Casals-Ruiz
Ilya Kazachkov

July 2011 • Volume 212 • Number 999 (end of volume) • ISSN 0065-9266

American Mathematical Society
Providence, Rhode Island

Library of Congress Cataloging-in-Publication Data

Casals-Ruiz, Montserrat.

On systems of equations over free partially commutative groups / Montserrat Casals-Ruiz, Ilya Kazachkov.
    p. cm. — (Memoirs of the American Mathematical Society, ISSN 0065-9266 ; no. 999)
    "Volume 212, number 999 (end of volume)."
    Includes bibliographical references and index.
    ISBN 978-0-8218-5258-3 (alk. paper)
    1. Equations, Abelian. 2. Abelian groups. I. Kazachkov, Ilya. II. Title.
QA215.C37   2011
512'.25—dc22
                                                                                                    2011011927

Memoirs of the American Mathematical Society

This journal is devoted entirely to research in pure and applied mathematics.

Publisher Item Identifier. The Publisher Item Identifier (PII) appears as a footnote on the Abstract page of each article. This alphanumeric string of characters uniquely identifies each article and can be used for future cataloguing, searching, and electronic retrieval.

Subscription information. Beginning with the January 2010 issue, *Memoirs* is accessible from www.ams.org/journals. The 2011 subscription begins with volume 209 and consists of six mailings, each containing one or more numbers. Subscription prices are as follows: for paper delivery, US$741 list, US$592.80 institutional member; for electronic delivery, US$667 list, US$533.60 institutional member. Upon request, subscribers to paper delivery of this journal are also entitled to receive electronic delivery. If ordering the paper version, subscribers outside the United States and India must pay a postage surcharge of US$69; subscribers in India must pay a postage surcharge of US$95. Expedited delivery to destinations in North America US$58; elsewhere US$167. Subscription renewals are subject to late fees. See www.ams.org/help-faq for more journal subscription information. Each number may be ordered separately; *please specify number* when ordering an individual number.

Back number information. For back issues see www.ams.org/bookstore.

Subscriptions and orders should be addressed to the American Mathematical Society, P. O. Box 845904, Boston, MA 02284-5904 USA. *All orders must be accompanied by payment.* Other correspondence should be addressed to 201 Charles Street, Providence, RI 02904-2294 USA.

Copying and reprinting. Individual readers of this publication, and nonprofit libraries acting for them, are permitted to make fair use of the material, such as to copy a chapter for use in teaching or research. Permission is granted to quote brief passages from this publication in reviews, provided the customary acknowledgment of the source is given.

Republication, systematic copying, or multiple reproduction of any material in this publication is permitted only under license from the American Mathematical Society. Requests for such permission should be addressed to the Acquisitions Department, American Mathematical Society, 201 Charles Street, Providence, Rhode Island 02904-2294 USA. Requests can also be made by e-mail to reprint-permission@ams.org.

Memoirs of the American Mathematical Society (ISSN 0065-9266) is published bimonthly (each volume consisting usually of more than one number) by the American Mathematical Society at 201 Charles Street, Providence, RI 02904-2294 USA. Periodicals postage paid at Providence, RI. Postmaster: Send address changes to Memoirs, American Mathematical Society, 201 Charles Street, Providence, RI 02904-2294 USA.

© 2010 by the American Mathematical Society. All rights reserved.
Copyright of individual articles may revert to the public domain 28 years
after publication. Contact the AMS for copyright status of individual articles.
This publication is indexed in *Science Citation Index*®, *SciSearch*®, *Research Alert*®,
CompuMath Citation Index®, *Current Contents*®/*Physical, Chemical & Earth Sciences*.
Printed in the United States of America.

∞ The paper used in this book is acid-free and falls within the guidelines
established to ensure permanence and durability.
Visit the AMS home page at http://www.ams.org/

10 9 8 7 6 5 4 3 2 1   16 15 14 13 12 11

Contents

List of Figures   vii

Chapter 1. Introduction   1

Chapter 2. Preliminaries   13
   2.1. Graphs and relations   13
   2.2. Elements of algebraic geometry over groups   13
   2.3. Formulas in the languages $\mathcal{L}_A$ and $\mathcal{L}_G$ 15
   2.4. First order logic and algebraic geometry   17
   2.5. Partially commutative groups   17
   2.6. Partially commutative monoids and DM-normal forms   22

Chapter 3. Reducing systems of equations over $\mathbb{G}$ to constrained generalised equations over $\mathbb{F}$ 25
   3.1. Definition of (constrained) generalised equations   27
   3.2. Reduction to generalised equations: from partially commutative groups to monoids   32
   3.3. Reduction to generalised equations: from partially commutative monoids to free monoids   37
   3.4. Example   41

Chapter 4. The process: construction of the tree T 49
   4.1. Preliminary definitions   50
   4.2. Elementary transformations   51
   4.3. Derived transformations   56
   4.4. Construction of the tree $T(\Omega)$ 63

Chapter 5. Minimal solutions   71

Chapter 6. Periodic structures   77
   6.1. Periodic structures   77
   6.2. Example   86
   6.3. Strongly singular and singular cases   88
   6.4. Example   91
   6.5. Regular case   92

Chapter 7. The finite tree $T_0(\Omega)$ and minimal solutions   99
   7.1. Automorphisms   100
   7.2. The finite subtree $T_0(\Omega)$ 103
   7.3. Paths $\mathfrak{p}(H)$ are in $T_0(\Omega)$ 110

Chapter 8. From the coordinate group $\mathbb{G}_{R(\Omega^*)}$ to proper quotients: the decomposition tree T_{dec} and the extension tree T_{ext} 121
8.1. The decomposition tree $T_{\text{dec}}(\Omega)$ 121
8.2. Example   127
8.3. The extension tree $T_{\text{ext}}(\Omega)$ 127

Chapter 9. The solution tree $T_{\text{sol}}(\Omega)$ and the main theorem   131
9.1. Example   136

Bibliography   139

Index   145

Glossary of notation   149

Abstract

Using an analogue of Makanin-Razborov diagrams, we give an effective description of the solution set of systems of equations over a partially commutative group (right-angled Artin group) $\mathbb{G}$. Equivalently, we give a parametrisation of $\mathrm{Hom}(G, \mathbb{G})$, where G is a finitely generated group.

Received by the editor October 24, 2008.
Article electronically published on December 22, 2010; S 0065-9266(2010)00628-8.
2010 *Mathematics Subject Classification*. Primary 20F70, Secondary 20F10; 20F36.
Key words and phrases. Equations in groups, partially commutative group, right-angled Artin group, Makanin-Razborov diagrams, algebraic geometry over groups.
The first author is supported by Programa de Formación de Investigadores del Departamento de Educación, Universidades e Investigación del Gobierno Vasco.
The second author is supported by the Richard. H. Tomlinson Fellowship.

©2010 American Mathematical Society

List of Figures

1	Cancellation in a product of geodesic words $w_1w_2w_3w_4w_5 = 1$.	20
1	The commutation graph of $\mathbb{G}$.	42
2	Cancellation scheme for the solution $x = bac$, $y = c^{-1}a^{-1}d$, $z = e$ of S.	42
3	The generalised equation $\Omega'_{\mathcal{T}}$.	44
1	Elementary transformation ET 1: Cutting a base.	52
2	Elementary transformation ET 2: Transferring a base.	53
3	Elementary transformation ET 3: Removing a pair of matched bases.	53
4	Elementary transformation ET 4: Removing a linear base.	54
5	Elementary transformation ET 5: Introducing a new boundary.	55
6	Derived transformation D 1: Closing a section.	57
7	Derived transformation D 2: Transporting a closed section.	57
8	Cases 3-4: Moving constant bases.	65
9	Cases 5-6: Removing a pair of matched bases and free variables.	65
10	Cases 7-8: Linear variables.	66
11	Cases 9-10: Linear case.	66
12	Case 12: Quadratic case, entire transformation.	67
1	The graphs Γ, Γ_0, T_0 and T.	87
1	The cancellation scheme of H^+.	128
2	The generalised equations $\Upsilon_{\mathcal{T}}$, $\check{\Upsilon}$ and $\Omega(\mathcal{P}, R, \mathfrak{c}, \mathcal{T})$.	129
1	The graphs Π_v and $\mathcal{F}$.	137
2	Constructing the group G by the graph $\varphi_{v,6}(\Pi_v)$.	138
3	The commutation graph of $\mathbb{G}_{w_6}$.	138

CHAPTER 1

Introduction

> "Divide et impera"
> Principle of government
> of Roman Senate

Equations are one of the most natural algebraic objects; they have been studied at all times and in all branches of mathematics and lie at its core. Though the first equations to be considered were equations over integers, now equations are considered over a multitude of other algebraic structures: rational numbers, complex numbers, fields, rings, etc. Since equations have always been an extremely important tool for a majority of branches of mathematics, their study developed into a separate, now classical, domain - algebraic geometry.

The algebraic structures we work with in this paper are groups. We would like to mention some important results in the area that established the necessary basics and techniques and that motivate the questions addressed in this paper. Though this historical introduction is several pages long, it is by no means complete and does not give a detailed account of the subject. We refer the reader to [**LS77**] and to [**Lyn80**] and references there for an extensive survey of the early history of equations in groups.

It is hard to date the beginning of the theory of equations over groups, since even the classical word and conjugacy problems formulated by Max Dehn in 1911 can be interpreted as the compatibility problem for very simple equations. Although the study of equations over groups now goes back almost a century, the foundations of a general theory of algebraic geometry over groups, analogous to the classical theory over commutative rings, were only set down relatively recently by G. Baumslag, A. Miasnikov and V. Remeslennikov, [**BMR99**].

Given a system of equations, the two main questions one can ask are whether the system is compatible and whether one can describe its set of solutions. It is these two questions that we now address.

Nilpotent and solvable groups. The solution of these problems is well-known for finitely generated abelian groups. V. Roman'kov showed that the compatibility problem is undecidable for nilpotent groups, see [**Rom77**]. Furthermore, in [**Rep83**] N. Repin refined the result of Roman'kov and showed that there exists a finitely presented nilpotent group for which the problem of recognising whether equations in one variable have solutions is undecidable. Later, in [**Rep84**], the author showed that for every non-abelian free nilpotent group of sufficiently large nilpotency class the compatibility problem for one-variable equations in undecidable. The compatibility problem is also undecidable for free metabelian groups, see [**Rom79b**].

Equations from the viewpoint of first-order logic are simply atomic formulas. Therefore, recognising if a system of equations and inequations over a group has a solution is equivalent to the decidability of the universal theory (or, which is equivalent, the existential theory) of this group. This is one of the reasons why often these two problems are intimately related. In general, both the compatibility problem and the problem of decidability of the universal theory are very deep. For instance, in [**Rom79a**] Roman'kov showed that the decidability of the universal theory of a free nilpotent group of class 2 is equivalent to the decidability of the universal theory of the field of rational numbers - a long-standing problem which, in turn, is equivalent to the Diophantine problem over rational numbers. For solvable groups, O. Chapuis in [**Ch98**] shows that if the universal theory of a free solvable group of class greater than or equal to 3 is decidable then so is the Diophantine problem over rational numbers.

Free groups and generalisations. In the case of free groups, both the compatibility problem and the problem of describing the solution set were long-standing problems. In this direction, works of R. Lyndon and A. Malcev, which are precursors to the solution of these problems, are of special relevance.

One-variable equations. One of the first types of equations to be considered was one-variable equations. In [**Lyn60**] R. Lyndon solved the compatibility problem and gave a description of the set of all solutions of a single equation in one variable (over a free group). Lyndon proved that the set of solutions of a single equation can be defined by a finite system of "parametric words". These parametric words were complicated and the number of parameters on which they depended was restricted only by the type of each equation considered. Further results were obtained by K. Appel [**Ap68**] and A.Lorents [**Lor63**], [**Lor68**], who gave the exact form of the parametric words, and Lorents extended the results to systems of equations with one variable. Unfortunately Appel's published proof has a gap, see p.87 of [**B74**], and Lorents announced his results without proof. In the year 2000, I. Chiswell and V. Remeslennikov gave a complete argument, see [**CR00**]. Instead of giving a description in terms of parametric words, they described the structure of coordinate groups of irreducible algebraic sets (in terms of their JSJ-decompositions) and, thereby, using the basic theory of algebraic geometry over free groups, they obtained a description of the solution set (viewed as the set of homomorphisms from the coordinate group to a free group). Recently, D. Bormotov, R. Gilman and A. Miasnikov in their paper [**BGM06**], gave a polynomial time algorithm that produces a description of the solution set of one-variable equations.

The parametric description of solutions of one-variable equations gave rise to a conjecture that the solution set of any system of equations could be described by a finite system of parametric words. In 1968, [**Ap68**], Appel showed that there are equations in three and more variables that can not be defined by a finite set of parametric words. Therefore, the method of describing the solution set in terms of parametric words was shown to be limited.

Two-variable equations. In an attempt to generalise the results obtained for one-variable equations, a more general approach, involving parametric functions, was suggested by Yu. Hmelevskiĭ. In his papers [**Hm71**], [**Hm72**], he gave a description of the solution set and proved the decidability of the compatibility problem of some systems of equations in two variables over a free group. This approach was later developed by Yu. Ozhigov in [**Oj83**], who gave a description of the solution set and proved the decidability of the compatibility problem for arbitrary equations in two variables.

However, it turned out that this method was not general either. In [**Raz84**], A. Razborov showed that there are equations whose set of solutions cannot be represented by a superposition of a finite number of parametric functions.

Recently, in [**T08**], N. Touikan, using the approach developed by Chiswell and Remeslennikov for one-variable equations, gave a description (in terms of the JSJ-decomposition) of coordinate groups of irreducible algebraic sets defined by a system of equations in two variables.

Quadratic equations. Because of their connections to surfaces and automorphisms thereof, quadratic equations have been studied since the beginning of the theory of equations over groups.

The first quadratic equation to be studied was the commutator equation $[x,y] = [a,b]$, see [**Niel**]. A description of the solution set of this equation was given in [**Mal62**] by A. Malcev in terms of parametric words in automorphisms and minimal solutions (with respect to these automorphisms).

Malcev's powerful idea was later developed by L. Commerford and C. Edmunds, see [**ComEd89**], and by R. Grigorchuk and P. Kurchanov, see [**GK89**], into a general method of describing the set of solutions of all quadratic equations over a free group. A more geometric approach to this problem was given by M. Culler, see [**Cul81**] and A. Olshanskii [**Ol89**].

Quadratic equations and Malcev's idea to use automorphisms and minimal solutions play a key role in the modern approach to describing the solution set of arbitrary systems of equations over free groups.

Because of their importance, quadratic equations were considered over other groups. The decidability of compatibility problem for quadratic equations over one relator free product of groups was proved by A. Duncan in [**Dun07**] and over certain small cancellation groups by Commerford in [**Com81**]. In [**Lys88**], I. Lysënok reduces the description of solutions of quadratic equations over certain small cancellation groups to the description of solutions of quadratic equations in free groups. Later, Grigorchuk and Lysënok gave a description of the solution set of quadratic equations over hyperbolic groups, see [**GrL92**].

Arbitrary systems of equations. A major breakthrough in the area was made by G. Makanin in his papers [**Mak77**] and [**Mak82**]. In his work, Makanin devised a process for deciding whether or not an arbitrary system of equations S over a free monoid (over a free group) is compatible. Later, using similar techniques, Makanin proved that the universal theory of a free group is decidable, see [**Mak84**]. Makanin's result on decidability of the universal theory of a free group together with an important result of Yu. Merzlyakov [**Mer66**] on quantifier elimination for positive formulae over free groups proves that the positive theory of a free group is decidable.

Makanin's ideas were developed in many directions. Remarkable progress was made by A. Razborov. In his work [**Raz85**], [**Raz87**], Razborov refined Makanin's process and used automorphisms and minimal solutions to give a complete description of the set of solutions of an arbitrary system of equations over a free group in terms of, what is now called, Makanin-Razborov diagrams (or Hom-diagrams). In their work [**KhM98b**], O. Kharlampovich and A. Miasnikov gave an important insight into Razborov's process and provided algebraic semantics for it. Using the process and having understanding of radicals of quadratic equations, the authors showed that the solution set of a system of equations can be canonically represented by the union of solution sets of a finite family of NTQ systems and gave an effective embedding of finitely generated fully residually free groups into coordinate groups of NTQ systems (ω-residually free towers), thereby giving a characterisation of such groups. Then, using Bass-Serre theory, they proved that finitely generated fully residually free groups are finitely presented and that one can effectively find a cyclic splitting of such groups. Analogous results were proved by Z. Sela using geometric techniques, see [**Sela01**]. Later, Kharlampovich and Myasnikov in [**KhM05a**] and Sela in [**Sela01**], developed Makanin-Razborov diagrams for systems of equations with parameters over a free group. These Makanin-Razborov diagrams encode the Makanin-Razborov diagrams of the systems of equations associated with each specialisation of the parameters. This construction plays a key role in a generalisation of Merzlyakov's theorem, in other words, in the proof of existence of Skolem functions for certain types of formulae (for NTQ systems of equations), see [**KhM05a**], [**Sela03**].

These results are an important piece of the solution of two well-known problems formulated by A. Tarski around 1945: the first of them states that the elementary theories of non-abelian free groups of different ranks coincide; and the second one states that the elementary theory of a free group is decidable. These problems were recently solved by O. Kharlampovich and A. Miasnikov in [**KhM06**] and the first one was independently solved by Z. Sela in [**Sela06**].

In another direction, Makanin's result (on decidability of equations over a free monoid) was developed by K.Schulz, see [**Sch90**], who proved that the compatibility problem of equations with regular constraints over a free monoid is decidable. Later V. Diekert, C. Gutiérrez and C. Hagenah, see [**DGH01**], reduced the compatibility problem of systems of equations over a free group with rational constraints to compatibility problem of equations with regular constraints over a free monoid.

Since then, one of the most successful methods of proving the decidability of the compatibility problem for groups (monoids) has been to reduce it to the compatibility problem over a free group (monoid) with constraints. The reduction of compatibility problem for torsion-free hyperbolic groups to free groups was made by E. Rips and Z. Sela in [**RS95**]; for relatively hyperbolic groups with virtually abelian parabolic subgroups by F. Dahmani in [**Dahm09**]; for partially commutative monoids by Yu. Matiasevich in [**Mat97**] (see also [**DMM99**]); for partially commutative groups by V. Diekert and A. Muscholl in [**DM06**]; for graph products of groups by V. Diekert and M. Lohrey in [**DL04**]; and for HNN-extensions with finite associated subgroups and for amalgamated products with finite amalgamated subgroups by M. Lohrey and G. Sénizergues in [**LS08**].

The complexity of Makanin's algorithm has received a lot of attention. The best result about arbitrary systems of equations over monoids is due to W. Plandowski. In a series of two papers [**Pl99a, Pl99b**] he gave a new approach to the compatibility problem of systems of equations over a free monoid and showed that this problem is in PSPACE. This approach was further extended by Diekert, Gutiérrez and Hagenah, see [**DGH01**] to systems of equations over free groups. Recently, O. Kharlampovich, I. Lysënok, A. Myasnikov and N. Touikan have shown that solving quadratic equations over free groups is NP-complete, [**KhLMT08**].

Another important development of the ideas of Makanin is due to E. Rips and is now known as the Rips' machine. In his work, Rips interprets Makanin's algorithm in terms of partial isometries of real intervals, which leads him to a classification theorem of finitely generated groups that act freely on $\mathbb{R}$-trees. A complete proof of Rips' theorem was given by D. Gaboriau, G. Levitt, and F. Paulin, see [**GLP94**], and, independently, by M. Bestvina and M. Feighn, see [**BF95**], who also generalised Rips' result to give a classification theorem of groups that have a stable action on $\mathbb{R}$-trees. Recently, F. Dahmani and V. Guirardel proved the decidability of the compatibility problem for virtually free groups generalising Rips' machine, see [**DG09**].

The existence of analogues of Makanin-Razborov diagrams has been proven for different groups. In [**Sela02**], for torsion-free hyperbolic groups by Z. Sela; in [**Gr05**], for torsion-free relatively hyperbolic groups relative to free abelian subgroups by D. Groves; for fully residually free (or limit) groups in [**KhM05b**], by O. Kharlampovich and A. Miasnikov and in [**Al07**], by E. Alibegović; for free products of groups in [**JS09**] by E. Jaligot and Z. Sela and in [**CK09**] by M. Casals-Ruiz and I. Kazachkov.

Other results. Other well-known problems and results were studied in relation to equations in groups. In analogy to Galois theory, the problem of adjoining roots of equations was considered in the first half of the twentieth century leading to the classical construction of an HNN-extension of a group G introduced by B. Higman, B. Neumann and H. Neumann in [**HNN49**] as a construction of an extension of a group G in which the "conjugacy" equation $x^{-1}gx = g'$ is compatible. Another example is the characterisation of the structure of elements g for which an equation of the form $w = g$ is compatible, also known as the endomorphism problem. A well known result in this direction is the solution of the endomorphism problem for the commutator equation $[x, y] = g$ in a free group obtained in [**Wic62**] by M. Wicks in terms of, what now are called, Wicks forms. Wicks forms were generalised to free products of groups and to higher genus by A. Vdovina in [**Vd97**], to hyperbolic groups by S. Fulthorp in [**Ful04**] and to partially commutative groups by S. Shestakov, see [**Sh05**]. The endomorphism problem for many other types of equations over free groups was studied by P. Schupp, see [**Sch69**], C. Edmunds, see [**Edm75**], [**Edm77**] and was generalised to free products of groups by L. Comerford and C. Edmunds, see [**ComEd81**].

We now focus on equations over partially commutative groups. Recall that a partially commutative group $\mathbb{G}$ is a group given by a presentation of the form $\langle a_1, \ldots, a_\mathfrak{r} \mid R \rangle$, where R is a subset of the set $\{[a_i, a_j] \mid i, j = 1, \ldots, \mathfrak{r}, i \neq j\}$. This means that the only relations imposed on the generators are commutation of some of the generators. In particular, both free abelian groups and free groups are partially commutative groups.

As we have mentioned above, given a system of equations, the two main questions one can ask are whether the system is compatible and whether one can describe its set of solutions. It is known that the compatibility problem for systems of equations over partially commutative groups is decidable, see [**DM06**]. Moreover, the universal (existential) and positive theories of partially commutative groups are also decidable, see [**DL04**] and [**CK07**]. But, on the other hand, until now there were not even any partial results known about the description of the solution sets of systems of equations over partially commutative groups.

Nevertheless, in the case of partially commutative groups, other problems, involving particular equations, have been studied. We would like to mention two papers by S. Shestakov, see [**Sh05**] and [**Sh06**], where the author solves the endomorphism problem for the equations $[x, y] = g$ and $x^2 y^2 = g$ correspondingly, and a result of J. Crisp and B. Wiest from [**CW04**] stating that partially commutative groups satisfy Vaught's conjecture, i.e. that if a tuple (X, Y, Z) of elements from $\mathbb{G}$ is a solution of the equation $x^2 y^2 z^2 = 1$, then X, Y and Z pairwise commute.

In this paper, we effectively construct an analogue of Makanin-Razborov diagrams and use it to give a description of the solution set of systems of equations over partially commutative groups. It seems to the authors that the importance of the work presented in this paper lies not only in the construction itself but in the fact that it enables consideration of analogues of the (numerous) consequences of the classical Makanin-Razborov process in more general circumstances.

The classes of groups for which Makanin-Razborov diagrams have been constructed so far are generalisations of free groups with a common feature: all of them are CSA-groups (see Lemma 2.5 in [**Gr08**]). Recall that a group is called CSA if every maximal abelian subgroup is malnormal, see [**MR96**], or, equivalently, a group is CSA if every non-trivial abelian subgroup is contained in a unique maximal abelian subgroup. In Proposition 9 in [**MR96**], it is proved that if a group is CSA, then it is commutative transitive (the commutativity relation is transitive) and thus the centralisers of its (non-trivial) elements are abelian, it is directly indecomposable, has no non-abelian normal subgroups, has trivial centre, has no non-abelian solvable subgroups and has no non-abelian Baumslag-Solitar subgroups. This shows that the CSA property imposes strong restrictions on the structure of the group and, especially, on the structure of the centralisers of its elements. Therefore, numerous classes of groups, even geometric, are not CSA.

The CSA property is important in the constructions of analogues of Makanin-Razborov diagrams constructed before. The fact that partially commutative groups are *not* CSA, conceptually, shows that this property is not essential for constructing Makanin-Razborov diagrams and that the approach developed in this paper opens the possibility for other groups to be taken in consideration: graph products of groups, particular types of HNN-extensions (amalgamated products), partially commutative-by-finite, fully residually partially commutative groups, particular types of small cancellation groups, torsion-free relatively hyperbolic groups, some torsion-free CAT(0) groups and more.

On the other hand, as mentioned above, Schulz proved that Makanin's process to decide the compatibility problem of equations carries over to systems of equations over a free monoid with regular constraints. In our construction, we show that Razborov's results can be generalised to systems of equations (over a free monoid) with constraints characterised by certain axioms. Therefore, two natural problems arise: to understand for which classes of groups the description of solutions of systems of equations reduces to this setting, and to understand which other constraints can be considered in the construction of Makanin-Razborov diagrams.

In another direction, the Makanin-Razborov process is one of the main ingredients in the effective construction of the JSJ-decomposition for fully residually free groups and in the characterisation of finitely generated fully residually free groups as given by Kharlampovich and Myasnikov in [**KhM05b**] and [**KhM98b**], correspondingly. Therefore, the process we construct in this paper may be one of the techniques one can use to understand the structure of finitely generated fully residually partially commutative groups, or, which in this case is equivalent, of finitely generated residually partially commutative groups, and thus, in particular, the structure of all finitely generated residually free groups.

In the case of free groups, Rips' machine leads to a classification theorem of finitely generated groups that have a stable action on $\mathbb{R}$-trees. Therefore, our process may be useful for understanding finitely generated groups that act nicely on certain CAT(0) spaces.

The structure of subgroups of partially commutative groups is very complex, see [**HW08**], and some of the subgroups exhibit odd finiteness properties, see [**BB97**]. Nevertheless, recent results on partially commutative groups suggest that the attempts to generalise some of the well-known results for free groups have been, at least to some extent, successful. We would like to mention here the recent progress of R. Charney and K. Vogtmann on the outer space of partially commutative groups, of M. Day on the generalisation of peak-reduction algorithm (Whitehead's theorem) for partially commutative groups, and of a number of authors on the structure of automorphism groups of partially commutative groups, see [**Day08a**], [**Day08b**], [**CCV07**], [**GPR07**], [**DKR08**]. This and the current paper makes us optimistic about the future of this emerging area.

To get a global vision of the long process we describe in this paper, we would like to begin by a brief comparison of the method of solving equations over a free monoid (as the reader will see the setting we reduce the study of systems of equations over partially commutative groups is rather similar to it) to Gaussian elimination in linear algebra. Though, technically very different, we want to point out that the general guidelines of both methods have a number of common features. Given a system of $\mathfrak{m}$ linear equations in k variables over a field K, the algorithm in linear algebra firstly encodes this system as an $\mathfrak{m} \times (k+1)$ matrix with entries in K. Then it uses elementary Gauss transformations of matrices (row permutation, multiplication of a row by a scalar, etc) in a particular order to take the matrix to a row-echelon form, and then it produces a description of the solution set of the system of linear equations with the associated matrix in row-echelon form. Furthermore, the algorithm has the following property: every solution of the (system of equations corresponding to the) matrix in row-echelon form gives rise to a solution of the original system and, conversely, every solution of the original system is induced by a solution of the (system of equations corresponding to the) matrix in row-echelon form. Let us compare this algorithm with its group-theoretic counterpart, a variation of which we present in this paper.

Given a system of equations over a free monoid, we introduce a combinatorial object - a generalised equation, and establish a correspondence between systems of equations over a free monoid and generalised equations (for the purposes of this introduction, the reader may think of generalised equations as just systems of equations over a monoid). Graphic representations of generalised equations will play the role of matrices in linear algebra.

We then define certain transformations of generalised equations, see Sections 4.2 and 4.3. One of the differences is that in the case of systems of linear equations, applying an elementary Gauss transformation to a matrix one obtains a *single* matrix. In our case, applying some of the (elementary and derived) transformations to a generalised equation one gets a finite family of generalised equations. Therefore, the method we describe here, results in an oriented rooted tree of generalised equations instead of a sequence of matrices.

A case-based algorithm, described in Section 4.4, applies transformations in a specific order to generalised equations. The branch of the algorithm terminates, when the generalised equation obtained is "simple" (there is a formal definition of being "simple"). The algorithm then produces a description of the set of solutions of "simple" generalised equations.

One of the main differences is that a branch of the procedure described may be *infinite*. Using particular automorphisms of coordinate groups of generalised equations as parameters, one can prove that a finite tree is sufficient to give a description of the solution set. Thus, the parametrisation of the solutions set will be given using a finite tree, recursive groups of automorphisms of finitely many coordinate groups of generalised equations and solutions of "simple" generalised equations.

We now briefly describe the content of each of the sections of the paper.

In Chapter 2 we establish the basics we will need throughout the paper.

The goal of Chapter 3 is to present the set of solutions of a system of equations over a partially commutative group as a union of sets of solutions of finitely many constrained generalised equations over a free monoid. The family of solutions of the collection constructed of generalised equations describes all solutions of the initial system over a partially commutative group. The term "constrained generalised equation over a free monoid (partially commutative monoid)" can be misleading, since their solutions are not tuples of elements from a free monoid (partially commutative monoid), but tuples of reduced elements of the partially commutative group, that satisfy the equalities in a free monoid (partially commutative monoid).

This reduction is performed in two steps. Firstly, we use an analogue of the notion of a cancellation tree for free groups to reduce the study of systems of equations over a partially commutative group to the study of constrained generalised equations over a partially commutative monoid. We show that van Kampen diagrams over partially commutative groups can be taken to a "standard form" and therefore the set of solutions of a given system of equations over a partially commutative group defines only finitely many types of van Kampen diagram in standard form, i.e. finitely many different cancellation schemes. For each of these cancellation schemes, we then construct a constrained generalised equation over a partially commutative monoid.

We further show that to a given generalised equation over a partially commutative monoid one can associate a finite collection of (constrained) generalised equations over a free monoid. The family of solutions of the generalised equations from this collection describes all solutions of the initial generalised equation over a partially commutative monoid. This reduction relies on the ideas of Yu. Matiyasevich, see [**Mat97**], V. Diekert and A. Muscholl, see [**DM06**], that state that there are only finitely many ways to take a product of words in the trace monoid (written in normal form) into normal form. We apply these results to reduce the study of solutions of generalised equations over a partially commutative monoid to the study of solutions of constrained generalised equations over a free monoid.

In Chapter 4 in order to describe the solution set of a constrained generalised equation over a free monoid we describe a branching rewriting process for constrained generalised equations. For a given generalised equation, this branching process results in a locally finite, possibly infinite, oriented rooted tree. The vertices of this tree are labelled by (constrained) generalised equations over a free monoid. Its edges are labelled by epimorphisms of the corresponding coordinate groups. Moreover, for every solution H of the original generalised equation, there exists a path in the tree from the root vertex to another vertex v and a solution $H^{(v)}$ of the generalised equation corresponding to v such that the solution H is a composition of the epimorphisms corresponding to the edges in the tree and the solution $H^{(v)}$. Conversely, to any path from the root to a vertex v in the tree, and any solution $H^{(v)}$ of the generalised equation labelling v, there corresponds a solution of the original generalised equation.

The tree described is, in general, infinite. In Lemma 4.19 we give a characterisation of the three types of infinite branches that it may contain: namely linear, quadratic and general. The aim of the remainder of the paper is, basically, to define the automorphism groups of coordinate groups that are used in the parametrisation of the solution sets and to prove that, using these automorphisms all solutions can be described by a finite tree.

Chapter 5 and 6 contain technical results used in Chapter 7.

In Chapter 5 we use automorphisms to introduce a reflexive, transitive relation on the set of solutions of a generalised equation. We use this relation to introduce the notion of a minimal solution and describe the behaviour of minimal solutions with respect to transformations of generalised equations.

In Chapter 6 we introduce the notion of periodic structure. Informally, a periodic structure is an object one associates to a generalised equation that has a solution that contains subwords of the form P^k for arbitrary large k. The aim of this chapter is two-fold. On the one hand, to understand the structure of the coordinate group of such a generalised equation and to define a certain finitely generated subgroup of its automorphism group. On the other hand, to prove that, using automorphisms from the automorphism group described, either one can bound the exponent of periodicity k or one can obtain a solution of a proper generalised equation.

Chapter 7 contains the core of the proof of the main results. In this chapter we deal with the three types of infinite branches and construct a finite oriented rooted subtree T_0 of the infinite tree described above. This tree has the property that for every leaf either one can trivially describe the solution set of the generalised equation assigned to it, or from every solution of the generalised equation associated to it, one gets a solution of a proper generalised equation using automorphisms defined in Chapter 6.

The strategy for the linear and quadratic branches is common: firstly, using a combinatorial argument, we prove that in such infinite branches a generalised equation repeats thereby giving rise to an automorphism of the coordinate group. Then, we show that such automorphisms are contained in a recursive group of automorphisms. Finally, we prove that minimal solutions with respect to this recursive group of automorphisms factor through sub-branches of bounded height.

The treatment for the general branch is more complex. We show that using automorphisms defined for quadratic branches and automorphisms defined in Chapter 6, one can take any solution to a solution of a proper generalised equation or to a solution of bounded length.

In Section 8.1 we prove that the number of proper generalised equations through which the solutions of the leaves of the finite tree T_0 factor is finite and this allows us to extend T_0 and obtain a tree T_{dec} with the property that for every leaf either one can trivially describe the solution set of the generalised equation assigned to it, or the edge with an end in the leaf is labelled by a proper epimorphism. Since partially commutative groups are equationally Noetherian and thus any sequence of proper epimorphisms of coordinate groups is finite, an inductive argument, given in Section 8.3, shows that we can construct a tree T_{ext} with the property that for every leaf one can trivially describe the solution set of the generalised equation assigned to it.

In the last section we construct a tree T_{sol} as an extension of the tree T_{ext} with the property that for every leaf the coordinate group of the generalised equation associated to it is a fully residually $\mathbb{G}$ partially commutative group and one can trivially describe its solution set.

The following theorem summarises one of the main results of the paper.

THEOREM. *Let $\mathbb{G}$ be the free partially commutative group with the underlying commutation graph $\mathcal{G}$ and let G be a finitely generated ($\mathbb{G}$-)group. Then the set of all ($\mathbb{G}$-)homomorphisms $\operatorname{Hom}(G, \mathbb{G})$ ($\operatorname{Hom}_{\mathbb{G}}(G, \mathbb{G})$, correspondingly) from G to $\mathbb{G}$ can be effectively described by a finite rooted tree. This tree is oriented from the root, all its vertices except for the root and the leaves are labelled by coordinate groups of generalised equations. The leaves of the tree are labelled by $\mathbb{G}$-fully residually $\mathbb{G}$ partially commutative groups $\mathbb{G}_{w_i}$ (described in Chapter 9).*

Edges from the root vertex correspond to a finite number of ($\mathbb{G}$-)homomorphisms from G into coordinate groups of generalised equations. To each vertex group we assign a group of automorphisms. Each edge (except for the edges from the root and the edges to the final leaves) in this tree is labelled by an epimorphism, and all the epimorphisms are proper. Every ($\mathbb{G}$-)homomorphism from G to $\mathbb{G}$ can be written as a composition of the ($\mathbb{G}$-)homomorphisms corresponding to the edges, automorphisms of the groups assigned to the vertices, and a ($\mathbb{G}$-)homomorphism $\psi = (\psi_j)_{j \in J}$, $|J| \leq 2^{\mathfrak{r}}$ into $\mathbb{G}$, where $\mathfrak{r}$ is the number of canonical generators of $\mathbb{G}$, $\psi_j : \mathbb{H}_j[Y] \to \mathbb{H}_j$ and $\mathbb{H}_j$ is the free partially commutative subgroup of $\mathbb{G}$ defined by some full subgraph of $\mathcal{G}$.

A POSTERIORI REMARK. In his work on systems of equations over a free group [**Raz85**], [**Raz87**], A. Razborov uses a result of J. McCool on automorphism group of a free group (that the stabiliser of a finite set of words is finitely presented), see [**McC75**], to prove that the automorphism groups of the coordinate groups associated to the vertices of the tree T_{sol} are finitely generated. When this paper was already written M. Day published two preprints, [**Day08a**] and [**Day08b**] on automorphism groups of partially commutative groups. The authors believe that using the results of this paper and techniques developed by M. Day, one can prove that the automorphism groups used in this paper are also finitely generated.

Acknowledgement. We wish to thank Olga Kharlampovich for her unconditional support and her patience answering our questions. We are grateful to Alexei

Miasnikov for introducing us into this subject and for many stimulating discussions we have had. We thank the referee for his/her careful reading and suggestions that helped us to improve the exposition.

CHAPTER 2

Preliminaries

2.1. Graphs and relations

In this section we introduce notation for graphs that we use in this paper. Let $\Gamma = (V(\Gamma), E(\Gamma))$ be an oriented graph, where $V(\Gamma)$ is the set of vertices of Γ and $E(\Gamma)$ is the set of edges of Γ. If an edge $e : v \to v'$ has *origin* v and *terminus* v', we sometimes write $e = v \to v'$. We always denote the paths in a graph by letters $\mathfrak{p}$ and $\mathfrak{s}$, and cycles by the letter $\mathfrak{c}$. To indicate that a path $\mathfrak{p}$ begins at a vertex v and ends at a vertex v' we write $\mathfrak{p}(v, v')$. If not stated otherwise, we assume that all paths we consider are simple. For a path $\mathfrak{p}(v, v') = e_1 \ldots e_l$ by $(\mathfrak{p}(v, v'))^{-1}$ we denote the reverse (if it exists) of the path $\mathfrak{p}(v, v')$, that is a path $\mathfrak{p}'$ from v' to v, $\mathfrak{p}' = e_l^{-1} \ldots e_1^{-1}$, where e_i^{-1} is the inverse of the edge e_i, $i = 1, \ldots l$.

Usually, the edges of the graph are labelled by certain words or letters. The *label* of a path $\mathfrak{p} = e_1 \ldots e_l$ is the concatenation of labels of the edges $e_1 \ldots e_l$.

Let Γ be an oriented rooted tree, with the root at a vertex v_0 and such that for any vertex $v \in V(\Gamma)$ there exists a unique path $\mathfrak{p}(v_0, v)$ from v_0 to v. The length of this path from v_0 to v is called the *height of the vertex* v. The number $\max_{v \in V(\Gamma)} \{\text{height of } v\}$ is called the *height of the tree* Γ. We say that a vertex v of height l is *above* a vertex v' of height l' if and only if $l > l'$ and there exists a path of length $l - l'$ from v' to v.

Let S be an arbitrary finite set. Let $\Re$ be a symmetric binary relation on S, i.e. $\Re \subseteq S \times S$ and if $(s_1, s_2) \in \Re$ then $(s_2, s_1) \in \Re$. Let $s \in S$, then by $\Re(s)$ we denote the following set:

$$\Re(s) = \{s_1 \in S \mid \Re(s, s_1)\}.$$

2.2. Elements of algebraic geometry over groups

In [**BMR99**] G. Baumslag, A. Miasnikov and V. Remeslennikov lay down the foundations of algebraic geometry over groups and introduce group-theoretic counterparts of basic notions from algebraic geometry over fields. We now recall some of the basics of algebraic geometry over groups. We refer to [**BMR99**] for details.

Algebraic geometry over groups centers around the notion of a *G-group*, where G is a fixed group generated by a set A. These G-groups can be likened to algebras over a unitary commutative ring, more specially a field, with G playing the role of the coefficient ring. We therefore, shall consider the category of G-groups, i.e. groups which contain a designated subgroup isomorphic to the group G. If H and K are G-groups then a homomorphism $\varphi : H \to K$ is a *G-homomorphism* if $\varphi(g) = g$ for every $g \in G$. In the category of G-groups morphisms are G-homomorphisms;

subgroups are G-subgroups, etc. By $\mathrm{Hom}_G(H,K)$ we denote the set of all G-homomorphisms from H into K. A G-group H is termed *finitely generated G-group* if there exists a finite subset $C \subset H$ such that the set $G \cup C$ generates H. It is not hard to see that the free product $G * F(X)$ is a free object in the category of G-groups, where $F(X)$ is the free group with basis $X = \{x_1, x_2, \ldots, x_n\}$. This group is called the free G-group with basis X, and we denote it by $G[X]$.

For any element $s \in G[X]$ the formal equality $s = 1$ can be treated, in an obvious way, as an *equation* over G. In general, for a subset $S \subset G[X]$ the formal equality $S = 1$ can be treated as *a system of equations* over G with coefficients in A. In other words, every equation is a word in $(X \cup A)^{\pm 1}$. Elements from X are called *variables*, and elements from $A^{\pm 1}$ are called *coefficients* or *constants*. To emphasize this we sometimes write $S(X, A) = 1$.

A *solution* U of the system $S(X) = 1$ over a group G is a tuple of elements $g_1, \ldots, g_n \in G$ such that every equation from S vanishes at $(g_1, \ldots, g_n)$, i.e. $S_i(g_1, \ldots, g_n) = 1$ in G, for all $S_i \in S$. Equivalently, a solution U of the system $S = 1$ over G is a G-homomorphism $\pi_U : G[X] \to G$ induced by the map $\pi_U : x_i \mapsto g_i$ such that $S \subseteq \ker(\pi_U)$. When no confusion arises, we abuse the notation and write $U(w)$, where $w \in G[X]$, instead of $\pi_U(w)$.

Denote by $\mathrm{ncl}\langle S \rangle$ the normal closure of S in $G[X]$. Then every solution of $S(X) = 1$ in G gives rise to a G-homomorphism $G[X]/\mathrm{ncl}\langle S \rangle \to G$, and vice versa. The set of all solutions over G of the system $S = 1$ is denoted by $V_G(S)$ and is called the *algebraic set defined by S*.

For every system of equations S we set the *radical of the system S* to be the following subgroup of $G[X]$:

$$R(S) = \{T(X) \in G[X] \mid \forall g_1, \ldots, \forall g_n \, (S(g_1, \ldots, g_n) = 1 \to T(g_1, \ldots, g_n) = 1)\}.$$

It is easy to see that $R(S)$ is a normal subgroup of $G[X]$ that contains S. There is a one-to-one correspondence between algebraic sets $V_G(S)$ and radical subgroups $R(S)$ of $G[X]$. This correspondence is described in Lemma 2.1 below. Notice that if $V_G(S) = \emptyset$, then $R(S) = G[X]$.

It follows from the definition that

$$R(S) = \bigcap_{U \in V_G(S)} \ker(\pi_U).$$

In the lemma below we summarise the relations between radicals, systems of equations and algebraic sets. Note the similarity of these relations with the ones in algebraic geometry over fields, see [**Eis99**].

For a subset $Y \subseteq G^n$ define the *radical of Y* to be

$$R(Y) = \{T(X) \in G[X] \mid T(g_1, \ldots, g_n) = 1 \text{ for all } (g_1, \ldots, g_n) \in Y\}.$$

LEMMA 2.1.
(1) *The radical of a system $S \subseteq G[X]$ contains the normal closure $\mathrm{ncl}\langle S \rangle$ of S.*
(2) *Let Y_1 and Y_2 be subsets of G^n and S_1, S_2 subsets of $G[X]$. If $Y_1 \subseteq Y_2$ then $R(Y_1) \supseteq R(Y_2)$. If $S_1 \subseteq S_2$ then $R(S_1) \subseteq R(S_2)$.*
(3) *For any family of sets $\{Y_i \mid i \in I\}$, $Y_i \subseteq G^n$, we have*

$$R\left(\bigcup_{i \in I} Y_i\right) = \bigcap_{i \in I} R(Y_i).$$

(4) *A normal subgroup H of the group $G[X]$ is the radical of an algebraic set over G if and only if $R(V_G(H)) = H$.*
(5) *A set $Y \subseteq G^n$ is algebraic over G if and only if $V_G(R(Y)) = Y$.*
(6) *Let $Y_1, Y_2 \subseteq G^n$ be two algebraic sets, then*
$$Y_1 = Y_2 \text{ if and only if } R(Y_1) = R(Y_2).$$

Therefore the radical of an algebraic set describes it uniquely.

The quotient group
$$G_{R(S)} = G[X]/R(S)$$
is called the *coordinate group* of the algebraic set $V_G(S)$ (or of the system S). There exists a one-to-one correspondence between the algebraic sets and coordinate groups $G_{R(S)}$. More formally, the categories of algebraic sets and coordinate groups are equivalent, see Theorem 4, [**BMR99**].

A G-group H is called *G-equationally Noetherian* if every system $S(X) = 1$ with coefficients from G is equivalent over G to a finite subsystem $S_0 = 1$, where $S_0 \subset S$, i.e. the system S and its subsystem S_0 define the same algebraic set. If G is G-equationally Noetherian, then we say that G is equationally Noetherian. If G is equationally Noetherian then the Zariski topology over G^n is *Noetherian* for every n, i.e., every proper descending chain of closed sets in G^n is finite. This implies that every algebraic set V in G^n is a finite union of irreducible subsets, called *irreducible components* of V, and such a decomposition of V is unique. Recall that a closed subset V is *irreducible* if it is not a union of two proper closed (in the induced topology) subsets.

If $V_G(S) \subseteq G^n$ and $V_G(S') \subseteq G^m$ are algebraic sets, then a map $\phi : V_G(S) \to V_G(S')$ is a *morphism* of algebraic sets if there exist $f_1, \ldots, f_m \in G[x_1, \ldots, x_n]$ such that, for any $(g_1, \ldots, g_n) \in V_G(S)$,
$$\phi(g_1, \ldots, g_n) = (f_1(g_1, \ldots, g_n), \ldots, f_m(g_1, \ldots, g_n)) \in V_G(S').$$
Occasionally we refer to morphisms of algebraic sets as *word mappings*.

Algebraic sets $V_G(S)$ and $V_G(S')$ are called *isomorphic* if there exist morphisms $\psi : V_G(S) \to V_G(S')$ and $\phi : V_G(S') \to V_G(S)$ such that $\phi\psi = \mathrm{id}_{V_G(S)}$ and $\psi\phi = \mathrm{id}_{V_G(S')}$.

Two systems of equations S and S' over G are called *equivalent* if the algebraic sets $V_G(S)$ and $V_G(S')$ are isomorphic.

PROPOSITION 2.2. *Let G be a group and let $V_G(S)$ and $V_G(S')$ be two algebraic sets over G. Then the algebraic sets $V_G(S)$ and $V_G(S')$ are isomorphic if and only if the coordinate groups $G_R(S)$ and $G_{R(S')}$ are G-isomorphic.*

The notions of an equation, system of equation, solution of an equation, algebraic set defined by a system of equations, morphism between algebraic sets and equivalent systems of equations are categorical in nature. These notion carry over, in an obvious way, from the case of groups to the case of monoids.

2.3. Formulas in the languages $\mathcal{L}_A$ and $\mathcal{L}_G$

In this section we recall some basic notions of first-order logic and model theory. We refer the reader to [**ChKe73**] for details.

Let G be a group generated by the set A. The standard first-order language of group theory, which we denote by $\mathcal{L}$, consists of a symbol for multiplication $\cdot$, a symbol for inversion $^{-1}$, and a symbol for the identity 1. To deal with G-groups, we have to enrich the language $\mathcal{L}$ by all the elements from G as constants. In fact, as G is generated by A, it suffices to enrich the language $\mathcal{L}$ by the constants that correspond to elements of A, i.e. for every element of $a \in A$ we introduce a new constant c_a. We denote language $\mathcal{L}$ enriched by constants from A by $\mathcal{L}_A$, and by constants from G by $\mathcal{L}_G$. In this section we further consider only the language $\mathcal{L}_A$, but everything stated below carries over to the case of the language $\mathcal{L}_G$.

A group word in variables X and constants A is a word $W(X, A)$ in the alphabet $(X \cup A)^{\pm 1}$. One may consider the word $W(X, A)$ as a term in the language $\mathcal{L}_A$. Observe that every term in the language $\mathcal{L}_A$ is equivalent modulo the axioms of group theory to a group word in variables X and constants A. An *atomic formula* in the language $\mathcal{L}_A$ is a formula of the type $W(X, A) = 1$, where $W(X, A)$ is a group word in X and A. We interpret atomic formulas in $\mathcal{L}_A$ as equations over G, and vice versa. A *Boolean combination* of atomic formulas in the language $\mathcal{L}_A$ is a disjunction of conjunctions of atomic formulas and their negations. Thus every Boolean combination Φ of atomic formulas in $\mathcal{L}_A$ can be written in the form $\Phi = \bigvee_{i=1}^{m} \Psi_i$, where each Ψ_i has one of the following forms:

$$\bigwedge_{j=1}^{n}(S_j(X, A) = 1), \text{ or } \bigwedge_{j=1}^{n}(T_j(X, A) \neq 1),$$

$$\text{or } \bigwedge_{j=1}^{n}(S_j(X, A) = 1) \wedge \bigwedge_{k=1}^{m}(T_k(X, A) \neq 1).$$

It follows from general results on disjunctive normal forms in propositional logic that every quantifier-free formula in the language $\mathcal{L}_A$ is logically equivalent (modulo the axioms of group theory) to a Boolean combination of atomic ones. Moreover, every formula Φ in $\mathcal{L}_A$ with *free variables* $Z = \{z_1, \ldots, z_k\}$ is logically equivalent to a formula of the type

$$Q_1 x_1 Q_2 x_2 \ldots Q_l x_l \Psi(X, Z, A),$$

where $Q_i \in \{\forall, \exists\}$, and $\Psi(X, Z, A)$ is a Boolean combination of atomic formulas in variables from $X \cup Z$. Introducing fictitious quantifiers, if necessary, one can always rewrite the formula Φ in the form

$$\Phi(Z) = \forall x_1 \exists y_1 \ldots \forall x_k \exists y_k \Psi(x_1, y_1, \ldots, x_k, y_k, Z).$$

A first-order formula Φ is called a *sentence*, if Φ does not contain free variables.

A sentence Φ is called *universal* if and only if Φ is equivalent to a formula of the type:

$$\forall x_1 \forall x_2 \ldots \forall x_l \Psi(X, A),$$

where $\Psi(X, A)$ is a Boolean combination of atomic formulas in variables from X. We sometimes refer to universal sentences as to universal formulas.

A *quasi identity* in the language $\mathcal{L}_A$ is a universal formula of the form

$$\forall x_1 \cdots \forall x_l \left(\bigwedge_{i=1}^{m} r_i(X, A) = 1 \to s(X, A) = 1 \right),$$

where $r_i(X, A)$ and $S(X, A)$ are terms.

2.4. First order logic and algebraic geometry

The connection between algebraic geometry over groups and logic has been shown to be very deep and fruitful and, in particular, led to a solution of the Tarski's problems on the elementary theory of free group, see [**KhM06**], [**Sela06**].

In [**Rem89**] and [**MR00**] A. Myasnikov and V. Remeslennikov established relations between universal classes of groups, algebraic geometry and residual properties of groups, see Theorems 2.3 and 2.4 below. We refer the reader to [**MR00**] and [**Kaz07**] for proofs.

In order to state these theorems, we shall make us of the following notions. Let H and K be G-groups. We say that a family of G-homomorphisms $\mathcal{F} \subset \mathrm{Hom}_G(H, K)$ G-separates (G-discriminates) H into K if for every non-trivial element $h \in H$ (every finite set of non-trivial elements $H_0 \subset H$) there exists $\phi \in \mathcal{F}$ such that $h^\phi \neq 1$ ($h^\phi \neq 1$ for every $h \in H_0$). In this case we say that H is G-separated by K or that H is G-residually K (G-discriminated by K or that H is G-fully residually K). In the case that $G = 1$, we simply say that H is separated (discriminated) by K.

THEOREM 2.3. *Let G be an equationally Noetherian (G-)group. Then the following classes coincide:*

- *the class of all coordinate groups of algebraic sets over G (defined by systems of equations with coefficients in G);*
- *the class of all finitely generated (G-)groups that are (G-)separated by G;*
- *the class of all finitely generated (G-)groups that satisfy all the quasi-identities (in the language $\mathcal{L}_G$ (or $\mathcal{L}_A$)) that are satisfied by G;*
- *the class of all finitely generated (G-)groups from the (G-)prevariety generated by G.*

Furthermore, a coordinate group of an algebraic set $V_G(S)$ is (G-)separated by G by homomorphisms π_U, $U \in V_G(S)$, corresponding to solutions.

THEOREM 2.4. *Let G be an equationally Noetherian (G-)group. Then the following classes coincide:*

- *the class of all coordinate group of irreducible algebraic sets over G (defined by systems of equations with coefficients in G);*
- *the class of all finitely generated (G-)groups that are G-discriminated by G;*
- *the class of all finitely generated (G-)groups that satisfy all universal sentences (in the language $\mathcal{L}_G$ (or $\mathcal{L}_A$)) that are satisfied by G.*

Furthermore, a coordinate group of an irreducible algebraic set $V_G(S)$ is (G-)discriminated by G by homomorphisms π_U, $U \in V_G(S)$, corresponding to solutions.

2.5. Partially commutative groups

Partially commutative groups are widely studied in different branches of mathematics and computer science, which explains the variety of names they were given: *graph groups, right-angled Artin groups, semifree groups*, etc. Without trying to give an account of the literature and results in the field we refer the reader to a recent survey [**Char07**] and to the introduction and references in [**DKR07**].

Recall that a (free) *partially commutative* group is defined as follows. Let $\mathcal{G}$ be a finite, undirected, simplicial graph. Let $\mathcal{A} = V(\mathcal{G}) = \{a_1, \ldots, a_{\mathfrak{r}}\}$ be the set of

vertices of $\mathcal{G}$ and let $F(\mathcal{A})$ be the free group on $\mathcal{A}$. Let
$$R = \{[a_i, a_j] \in F(\mathcal{A}) \mid a_i, a_j \in \mathcal{A} \text{ and there is an edge of } \mathcal{G} \text{ joining } a_i \text{ to } a_j\}.$$
The partially commutative group corresponding to the (commutation) graph $\mathcal{G}$ is the group $\mathbb{G}(\mathcal{G})$ with presentation $\langle \mathcal{A} \mid R \rangle$. This means that the only relations imposed on the generators are commutation of some of the generators. When the underlying graph is clear from the context we write simply $\mathbb{G}$.

From now on $\mathcal{A} = \{a_1, \ldots, a_{\mathfrak{r}}\}$ always stands for a finite alphabet and its elements are called *letters*. We reserve the term *occurrence* to mean an occurrence of a letter or of its formal inverse in a word. In a more formal way, an occurrence is a pair (letter (or its inverse), its placeholder in the word).

Let $\mathcal{G}_1$ be a full subgraph of $\mathcal{G}$ and let $\mathcal{A}_1$ be its set of vertices. In [**EKR05**] it is shown that $\mathbb{G}(\mathcal{G}_1)$ is the subgroup of the group $\mathbb{G}$ generated by $\mathcal{A}_1$, $\mathbb{G}(\mathcal{G}_1) = \langle \mathcal{A}_1 \rangle$. Following [**DKR07**], we call $\mathbb{G}(\mathcal{G}_1) = \mathbb{G}(\mathcal{A}_1)$ a *canonical parabolic subgroup* of $\mathbb{G}$.

We denote the length of a word w by $|w|$. For a word $w \in \mathbb{G}$, we denote by $\overline{w}$ a *geodesic* of w. Naturally, $|\overline{w}|$ is called the length of an element $w \in \mathbb{G}$. An element $w \in \mathbb{G}$ is called *cyclically reduced* if the length of $\overline{w^2}$ is twice the length of $\overline{w}$ or, equivalently, the length of w is minimal in the conjugacy class of w.

Let u be a geodesic word in $\mathbb{G}$. We say that u is a left (right) *divisor* of an element w, $w \in \mathbb{G}$ if there exists a geodesic word $v \in \mathbb{G}$ and a geodesic word $\overline{w}$, $\overline{w} = w$, such that $\overline{w} = uv$ ($\overline{w} = vu$, respectively) and $|\overline{w}| = |u| + |v|$. In this case, we also say that u *left-divides* w (*right-divides* w, respectively). Let u, v be geodesic words in $\mathbb{G}$. If $|\overline{uv}| = |u| + |v|$ we sometimes write $u \circ v$ to stress that there is no cancellation between u and v.

For a given word w denote by $\mathrm{alph}(w)$ the set of letters occurring in w. For a word $w \in \mathbb{G}$ define $\mathbb{A}(w)$ to be the subgroup of $\mathbb{G}$ generated by all letters that do not occur in $\overline{w}$ and commute with w. The subgroup $\mathbb{A}(w)$ is well-defined (independent of the choice of a geodesic $\overline{w}$), see [**EKR05**]. Let $v, w \in \mathbb{G}$ be so that $[v, w] = 1$ and $\mathrm{alph}(v) \cap \mathrm{alph}(w) = \emptyset$, or, which is equivalent, $v \in \mathbb{A}(w)$ and $w \in \mathbb{A}(v)$. In this case we write $v \leftrightarrows w$.

Let $A, B \subseteq \mathbb{G}$ be arbitrary subsets of $\mathbb{G}$. We denote by $[A, B]$ the set $[A, B] = \{[a, b] \mid a \in A, b \in B\}$ (not to confuse with the more usual notation $[A, B]$ for the subgroup generated by the set $\{[a, b] \mid a \in A, b \in B\}$). Naturally, the notation $[A, B] = 1$ means that $[a, b] = 1$ for all $[a, b] \in [A, B]$. Analogously, given a set of words $W \subseteq \mathbb{G}$ we denote by $\mathrm{alph}(W)$ the set of letters that occur in a word $w \in W$, i.e. $\mathrm{alph}(W) = \bigcup_{w \in W} \mathrm{alph}(w)$, and by $\mathbb{A}(W)$ the intersection of the subgroups $\mathbb{A}(w)$ for all $w \in W$, i.e. $\mathbb{A}(W) = \bigcap_{w \in W} \mathbb{A}(w)$. Similarly, we write $A \leftrightarrows B$ whenever $\mathrm{alph}(A) \cap \mathrm{alph}(B) = \emptyset$ and $[A, B] = 1$.

We say that a word $w \in \mathbb{G}$ is written in the *lexicographical normal form* if the word w is geodesic and minimal with respect to the lexicographical ordering induced by the ordering
$$a_1 < a_2 \cdots < a_{\mathfrak{r}} < a_{\mathfrak{r}}^{-1} < \cdots < a_1^{-1}$$
of the set $\mathcal{A} \cup \mathcal{A}^{-1}$. Note that if w is written in the lexicographical normal form, the word w^{-1} is not necessarily written in the lexicographical normal form, see Remark 32, [**DM06**].

Henceforth, by the symbol '$\doteq$' we denote graphical equality of words, i.e. equality in the free monoid.

2.5. PARTIALLY COMMUTATIVE GROUPS

For a partially commutative group $\mathbb{G}$ consider its non-commutation graph Δ. The vertex set V of Δ is the set of generators $\mathcal{A}$ of $\mathbb{G}$. There is an edge connecting a_i and a_j if and only if $[a_i, a_j] \neq 1$. Note that the graph Δ is the complement graph the graph Γ. The graph Δ is a union of its connected components $I_1, \ldots, I_k$. In the above notation

$$(2.1) \qquad \mathbb{G} = \mathbb{G}(I_1) \times \cdots \times \mathbb{G}(I_k).$$

Consider $w \in \mathbb{G}$ and the set $\mathrm{alph}(w)$. For this set, just as above, consider the graph $\Delta(\mathrm{alph}(w))$ (it is a full subgraph of Δ). This graph can be either connected or not. If it is connected we will call w a *block*. If $\Delta(\mathrm{alph}(w))$ is not connected, then we can split w into the product of commuting words

$$(2.2) \qquad w = w_{j_1} \cdot w_{j_2} \cdots w_{j_t}; \; j_1, \ldots, j_t \in J,$$

where $|J|$ is the number of connected components of $\Delta(\mathrm{alph}(w))$ and the word w_{j_i} is a word in the letters from the j_i-th connected component. Clearly, the words $\{w_{j_1}, \ldots, w_{j_t}\}$ pairwise commute. Each word w_{j_i}, $i \in 1, \ldots, t$ is a block and so we refer to presentation (2.2) as the block decomposition of w.

An element $w \in \mathbb{G}$ is called a least root (or simply, root) of $v \in \mathbb{G}$ if there exists a positive integer $0 \neq m \in \mathbb{N}$ such that $v = w^m$ and there does not exists $w' \in \mathbb{G}$ and $0 \neq m' \in \mathbb{N}$ such that $w = w'^{m'}$. In this case we write $w = \sqrt{v}$. By a result from [**DK93**], partially commutative groups have least roots, that is the root element of v is defined uniquely.

The next result describes centralisers of elements in partially commutative groups.

THEOREM 2.5 (Centraliser Theorem, Theorem 3.10, [**Ser89**] and [**DK93**]). *Let $w \in \mathbb{G}$ be a cyclically reduced word and $w = v_1 \ldots v_k$ be its block decomposition. Then, the centraliser of w is the following subgroup of $\mathbb{G}$:*

$$C(w) = \langle \sqrt{v_1} \rangle \times \cdots \times \langle \sqrt{v_k} \rangle \times \mathbb{A}(w).$$

COROLLARY 2.6. *For any $w \in \mathbb{G}$ the centraliser $C(w)$ of w is an isolated subgroup of $\mathbb{G}$. In particular, $C(w) = C(\sqrt{w})$.*

In Section 3.1 we shall need an analogue of the definition of a cancellation scheme for a free group. We gave a description of cancellation schemes for partially commutative groups in [**CK07**]. We shall use the following result.

PROPOSITION 2.7 (Lemma 3.2, [**CK07**]). *Let $\mathbb{G}$ be a partially commutative group and let $w_1, \ldots w_k$ be geodesic words in $\mathbb{G}$ such that $w_1 \cdots w_k = 1$. Then, there exist geodesic words w_i^j, $1 \leq i, j \leq k$ such that for any $1 \leq l \leq k$ there exists the following geodesic presentation for w_l:*

$$w_l = w_l^{l-1} \cdots w_l^1 w_l^k \cdots w_l^{l+1},$$

where $w_l^i = w_i^{l\,-1}$.

This result is illustrated in Figure 1. In fact, Proposition 2.7 states that there exists a normal form for van Kampen diagrams over partially commutative groups and that, structurally, there are only finitely many possible normal forms of van Kampen diagrams for the product $w_1 \cdots w_k = 1$ corresponding to the different decompositions of the word w_i as a product of the *non-trivial* words w_i^j. We usually consider van Kampen diagrams in normal forms and think of them in terms of their structure. It is, therefore, natural to refer to van Kampen diagrams in normal forms as to *cancellation schemes*. We refer the reader to [**CK07**] for more details.

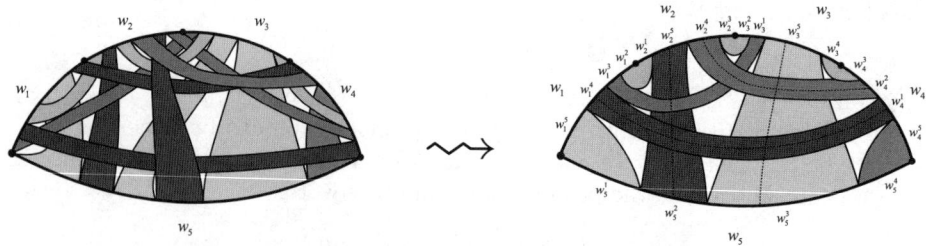

FIGURE 1. Cancellation in a product of geodesic words $w_1 w_2 w_3 w_4 w_5 = 1$.

Figure 1 (on the right), shows a cancellation scheme for the product of words $w_1 w_2 w_3 w_4 w_5 = 1$. All cancellation schemes for $w_1 w_2 w_3 w_4 w_5 = 1$ can be constructed from the one shown on Figure 1 (on the right) by making some of the bands trivial.

In the last section of this paper we use Proposition 2.12 to prove that certain partially commutative $\mathbb{G}$-groups are $\mathbb{G}$-fully residually $\mathbb{G}$. Proposition 2.12 is a generalisation of a result from [**CK07**] that states that if $\mathbb{G}$ is a non-abelian, directly indecomposable partially commutative group, then $\mathbb{G}[X]$ is $\mathbb{G}$-fully residually $\mathbb{G}$. The next two lemmas were used in the proof of this result in [**CK07**] and are necessary for our proof of Proposition 2.12.

LEMMA 2.8 (Lemma 4.11, [**CK07**]). *There exists an integer $N = N(\mathbb{G})$ such that the following statements hold.*

(1) *Let $b \in \mathbb{G}$ be a cyclically reduced block and let $z \in \mathbb{G}$ be so that b^{-1} does not left-divide and right-divide z. Then one has $b^{N+1} z b^{N+1} = b \circ b^N z b^N \circ b$.*
(2) *Let $b \in \mathbb{G}$ be a cyclically reduced block and let $z = z_1^{-1} z_2 z_1$, where z_2 is cyclically reduced. Suppose that b does not left-divide z, z^{-1}, z_2 and z_2^{-1}, and $[b, z] \neq 1$. Then one has $z^{b^{N+1}} = b \circ z^{b^N} \circ b^{-1}$.*

REMARK 2.9. In [**CK07**] we prove that N is a linear function of the centraliser dimension of the group $\mathbb{G}$, see [**DKR06**] for definition.

LEMMA 2.10 (Lemma 4.17, [**CK07**]). *Let $b \in \mathbb{G}$ be a cyclically reduced block element and let $w_1, w_2 \in \mathbb{G}$ be geodesic words of the form*

$$w_1 = b^{\delta_1} \circ g_1 \circ b^\epsilon, \quad w_2 = b^\epsilon \circ g_2 \circ b^{\delta_2}, \text{ where } \epsilon, \delta_1, \delta_2 = \pm 1.$$

Then the geodesic word $\overline{w_1 w_2}$ has the form $\overline{w_1 w_2} = b^{\delta_1} \circ w_3 \circ b^{\delta_2}$.

COROLLARY 2.11. *Let $\mathbb{G}$ be a partially commutative group and let $1 \neq w \in \mathbb{G}[x]$, $w = x^{k_1} g_1 \cdots g_{l-1} x^{k_l} g_l$, where $g_i, \in \mathbb{G}$, $g_1, \ldots, g_l \neq 1$. Suppose that there exists a cyclically reduced block element $b \in \mathbb{G}$ such that $[b, g_i] \neq 1$, $i = 1, \ldots, l-1$. Then there exists a positive integer N such that for all $n > N$ the homomorphism $\varphi_{b,n}$ induced by the map $x \mapsto b^n$ maps the word w to a non-trivial element of $\mathbb{G}$.*

PROOF. Let $g_i = g_{i,1}^{-1} g_{i,2} g_{i,1}$ be the cyclic decomposition of g_i. Taking b' to be a large enough power of the block element b, we may assume that b' does not left-divide the elements $g_{i,1}^{\pm 1}$, $g_{i,2}^{\pm 1}$, $i = 1, \ldots, l$.

Note that b' satisfies the assumptions of Lemma 2.8.

Consider the homomorphism $\varphi_{b,2n} : \mathbb{G}[x] \to \mathbb{G}$, defined by $x \mapsto b^{2n}$. Then

$$\varphi_{b,2n}(w) = b^{k_1 n} \left(b^{k_1 n} g_1 b^{k_2 n} \right) \cdot \left(b^{k_2 n} g_2 b^{k_2 n} \right) \cdots \left(b^{k_{l-1} n} g_{l-1} b^{k_l n} \right) \cdot b^{k_l n} g_l$$

By Lemma 2.8, there exists N such that if $n \geq N$, then every factor of $\varphi_{b,2n}(w)$ of the form $\left(b^{k_i n} g_i b^{k_{i+1} n} \right)$ has the form $b^{\text{sign}(k_i)} \circ \tilde{g}_i \circ b^{\text{sign}(k_{i+1})}$. The statement now follows from Lemma 2.10. □

PROPOSITION 2.12. *Let $\mathbb{G}$ be a partially commutative group and let $\mathbb{H}$ be a directly indecomposable canonical parabolic subgroup of $\mathbb{G}$. Then the group*

$$\mathbb{G}' = \langle \mathbb{G}, t \mid \text{rel}(\mathbb{G}), [t, C_{\mathbb{G}}(\mathbb{H})] = 1 \rangle,$$

where $C_{\mathbb{G}}(\mathbb{H})$ is the centraliser of $\mathbb{H}$ in $\mathbb{G}$, is a $\mathbb{G}$-discriminated by $\mathbb{G}$ partially commutative group.

PROOF. Firstly, note that by Theorem 2.5, the subgroup $C_{\mathbb{G}}(\mathbb{H})$ is a canonical parabolic subgroup of the group $\mathbb{G}$. It follows that the group $\mathbb{G}'$ is a partially commutative group.

Consider an element $g \in \mathbb{G}'$ written in the normal form induced by an ordering on $\mathcal{A}^{\pm 1} \cup \{t\}^{\pm 1}$, so that $t^{\pm 1}$ precedes any element $a \in \mathcal{A}^{\pm 1}$,

$$g = g_1 t^{k_1} \cdots g_l t^{k_l} g_{l+1},$$

where $[g_j, t] \neq 1$, $j = 1, \ldots, l$, $g_j \in \mathbb{G}$. Since $[g_j, t] \neq 1$, we get that $g_i \notin C_{\mathbb{G}}(\mathbb{H})$. Fix a block element b from $\mathbb{H}$, such that $[b, g_j] \neq 1$ for all $j = 1, \ldots, l$. Clearly, such b exists, since $\mathbb{H}$ is directly indecomposable. If we treat the word g as an element of $\mathbb{G}[t]$, then by Corollary 2.11, there exists a positive integer M such that for all $n \geq M$, the family of homomorphisms $\varphi_{b,n}$ induced by the map $t \mapsto b^n$, maps g to a non-trivial element of $\mathbb{G}$.

Since, by the choice of b, the element $\varphi_{b,n}(t)$ belongs to $\mathbb{H}$, it follows that the relations $[C_{\mathbb{G}}(\mathbb{H}), \varphi_{b,n}(t)] = 1$ are satisfied. Therefore, the family of homomorphisms $\{\varphi_{b,n}\}$ induces a $\mathbb{G}$-approximating family $\{\tilde{\varphi}_{b,n}\}$ of homomorphisms from the group $\mathbb{G}'$ to $\mathbb{G}$, see the diagram below.

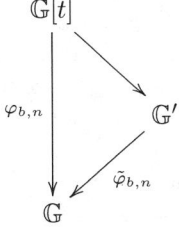

Considering, instead of one element g, a finite family of elements from $\mathbb{G}'$ and choosing the block element b in an analogous way, we get that the group $\mathbb{G}'$ is $\mathbb{G}$-discriminated by $\mathbb{G}$. □

The property of a group to be equationally Noetherian plays an important role in algebraic geometry over groups, see Theorems 2.3 and 2.4. It is known that every linear group (over a commutative, Noetherian, unitary ring) is equationally

Noetherian (see [**Guba86**], [**Br77**], [**BMR99**]). Partially commutative groups are linear, see [**Hum94**], hence equationally Noetherian.

In Chapter 9 we shall use the notion of a *graph product of groups*. The idea of a graph product, introduced in [**Gr90**], is a generalisation of the concept of a partially commutative group. Let $G_1, \ldots, G_k$, $G_i = \langle X_i \mid R_i \rangle$, $i = 1, \ldots, k$ be groups. Let $\mathcal{G}' = (V(\mathcal{G}'), E(\mathcal{G}'))$ be a finite, undirected, simplicial graph, $V(\mathcal{G}') = \{v_1, \ldots, v_k\}$.

A graph product $G = G_{\mathcal{G}'} = G_\Gamma(G_1, \ldots, G_k)$ of the groups $G_1, \ldots, G_k$ with respect to the graph $\mathcal{G}'$, is a group with a presentation of the form

$$\langle X_1, \ldots, X_k \mid R_1, \ldots, R_k, \mathcal{R} \rangle,$$

where $\mathcal{R} = \{[X_i, X_j] \mid v_i \text{ and } v_j \text{ are adjacent in } \mathcal{G}'\}$.

2.6. Partially commutative monoids and DM-normal forms

The aim of this section is to describe a normal form for elements of a partially commutative monoid. For our purposes, it is essential that the normal form be invariant with respect to inversion. As mentioned above, natural normal forms, such as the lexicographical normal form or the normal form arising from the bi-automatic structure on $\mathbb{G}$, see [**VW94**], do not have this property and hence can not be used. In [**DM06**] V. Diekert and A. Muscholl specially designed a normal form that has this property. We now define this normal form and refer the reader to [**DM06**] for details.

Let $\mathbb{G}$ be a partially commutative group given by the presentation $\langle \mathcal{A} \mid R \rangle$. Let $\mathbb{F} = \mathbb{F}(\mathcal{A}^{\pm 1})$ be the free monoid on the alphabet $\mathcal{A} \cup \mathcal{A}^{-1}$ and let $\mathbb{T} = \mathbb{T}(\mathcal{A}^{\pm 1})$ be the partially commutative monoid with involution given by the presentation:

$$\mathbb{T}(\mathcal{A}^{\pm 1}) = \langle \mathcal{A} \cup \mathcal{A}^{-1} \mid R_\mathbb{T} \rangle,$$

where $[a_i^\epsilon, a_j^\delta] \in R_\mathbb{T}$ if and only if $[a_i, a_j] \in R$, $\epsilon, \delta \in \{-1, 1\}$. The involution on $\mathbb{T}$ is induced by the operation of inversion in $\mathbb{G}$ and does not have fixed points. We refer to it as to the *inversion* in $\mathbb{T}$ and denote it by $^{-1}$.

Following [**DM06**], we call a maximal subset $C = \mathcal{C} \cup \mathcal{C}^{-1}$ of $\mathcal{A} \cup \mathcal{A}^{-1}$ such that $[a, c] \notin R_\mathbb{T}$ if and only if $[b, c] \notin R_\mathbb{T}$ for all $a, b \in C$ and $c \in \mathcal{A}^{\pm 1}$, a *clan*. A clan C is called *thin* if there exist $a \in C$ and $b \in \mathcal{A}^{\pm 1} \setminus C$ such that $[a, b] \in R_\mathbb{T}$ and is called *thick* otherwise. It follows that there is at most one thick clan and that the number of thin clans never equals 1.

Every element of $\mathcal{A} \cup \mathcal{A}^{-1}$ belongs to exactly one clan. If $\mathbb{T}$ is a direct product of d free monoids, then the number of thin clans is d for $d > 1$, and it is 0 for $d = 1$. In the following we pick a thin clan and we make it thick by removing commutation. It might be that the number of clans does not change, but the number of thin clans decreases. This is the reason why the definition of DM-normal form below is based on thin clans (instead of considering all clans).

It is convenient to encode an element of the partially commutative monoid as a finite labelled acyclic oriented graph $[V, E, \lambda]$, where V is the set of vertices, E is the set of edges and $\lambda : V \to \mathcal{A}^{\pm 1}$ is the labelling. Such a graph induces a labelled partial order $[V, E^*, \lambda]$. For an element $w \in \mathbb{T}$, $w = b_1 \cdots b_n$, $b_i \in \mathcal{A}^{\pm 1}$, we introduce the graph $[V, E, \lambda]$ as follows. The set of vertices of $[V, E, \lambda]$ is in one-to-one correspondence with the letters of w, $V = \{1, \ldots, n\}$. For the vertex j we set $\lambda(j) = b_j$. We define an edge from b_i to b_j if and only if both $i < j$ and $[b_i, b_j] \notin R_\mathbb{T}$. The graph $[V, E, \lambda]$ thereby obtained is called the *dependence graph* of

w. Up to isomorphism, the dependence graph of w is unique, and so is its induced labelled partial order, which we further denote by $[V, \leq, \lambda]$.

Let $c_1 < \cdots < c_q$ be the linearly ordered subset of $[V, \leq, \lambda]$ containing all vertices with label in the clan C. For the vertex $v \in V$, we define the *source point* $s(v)$ and and the *target point* $t(v)$ as follows:
$$s(v) = \sup\{i \mid c_i \leq v\}, \quad t(v) = \inf\{i \mid v \leq c_i\}.$$
By convention $\sup \emptyset = 0$ and $\inf \emptyset = q+1$. Thus, $0 \leq s(v) \leq q$, $1 \leq t(v) \leq q+1$ and $s(v) \leq t(v)$ for all $v \in V$. Note that we have $s(v) = t(v)$ if and only if the label of v belongs to C.

For $0 \leq s \leq t \leq q+1$, we define the *median position* $m(s,t)$. For $s = t$ we let $m(s,t) = s$. For $s < t$, by Lemma 1 in [**DM06**], there exist unique l and k such that $s \leq l < t$, $k \geq 0$ and
$$c_{s+1} \ldots c_l \in \mathbb{F}(\mathcal{C})(\mathcal{C}^{-1}\mathbb{F}(\mathcal{C}))^k, \quad c_{l+1} \ldots c_{t-1} \in (\mathbb{F}(\mathcal{C}^{-1})\mathcal{C})^k \mathbb{F}(\mathcal{C}^{-1}).$$
Then we define $m(s,t) = l + \frac{1}{2}$ and we call $m(s,t)$ the median position. Define the *global position* of $v \in V$ to be $g(v) = m(s(v), t(v))$.

We define the normal form $\operatorname{nf}(w)$ of an element $w \in \mathbb{T}$ by introducing new edges into the dependence graph $[V, E, \lambda]$ of w. Let $u, v \in V$ be such that $\lambda(v) \in C$ and $[\lambda(u), \lambda(v)] \in R_\mathbb{T}$. We define a new edge from u to v if $g(u) < g(v)$, otherwise we define a new edge from v to u. The new dependence graph $[V, \hat{E}, \lambda]$ defines a unique element of the trace monoid $\hat{\mathbb{T}}$, where $\hat{\mathbb{T}}$ is obtained from $\mathbb{T}$ by omitting the commutativity relations of the form $[c, a]$ for any $c \in C$ and any $a \in \mathcal{A}^{\pm 1}$. Note that the number of thin clans of $\hat{\mathbb{T}}$ is strictly less than the number of thin clans of $\mathbb{T}$. We proceed by designating a thin clan in $\hat{\mathbb{T}}$ and introducing new edges in the dependence graph $[V, \hat{E}, \lambda]$.

It is proved in Lemma 4, [**DM06**], that the normal form nf is a map from the trace monoid $\mathbb{T}$ to the free monoid $\mathbb{F}(\mathcal{A} \cup \mathcal{A}^{-1})$, which is compatible with inversion, i.e. it satisfies that $\pi(\operatorname{nf}(w)) = w$ and $\operatorname{nf}(w^{-1}) = \operatorname{nf}(w)^{-1}$, where $w \in \mathbb{T}$ and π is the canonical epimorphism from $\mathbb{F}(\mathcal{A} \cup \mathcal{A}^{-1})$ to $\mathbb{T}$.

We refer to this normal form as to the *DM-normal form* or simply as to the *normal form* of an element $w \in \mathbb{T}$.

CHAPTER 3

Reducing systems of equations over $\mathbb{G}$ to constrained generalised equations over $\mathbb{F}$

The notion of a generalised equation was introduced by Makanin in [**Mak77**]. A generalised equation is a combinatorial object which encodes a system of equations over a free monoid. In other words, given a system of equations S over a free monoid, one can construct a generalised equation $\Upsilon(S)$. Conversely, to a generalised equation Υ, one can canonically associate a system of equations S_Υ over the free monoid $\mathbb{F}$. The correspondence described has the following property. Given a system S the system $S_{\Upsilon(S)}$ is equivalent to S, i.e. the set of solutions defined by S and by $S_{\Upsilon(S)}$ are isomorphic; and vice versa, given a generalised equation Υ, one has that the generalised equations $\Upsilon(S_\Upsilon)$ and Υ are equivalent, see Definition 3.2 and Lemma 3.3.

The motivation for defining a generalised equation is two-fold. One the one hand, it gives an efficient way of encoding all the information about a system of equations and, on the other hand, elementary transformations, that are essential for Makanin's algorithm, see Section 4.2, have a cumbersome description in terms of systems of equations, but admit an intuitive one in terms of graphic representations of combinatorial generalised equations. In this sense graphic representations of generalised equations can be likened to matrices. In linear algebra there is a correspondence between systems of equations over a field k and matrices with elements from k. To describe the set of solutions of a system of equations, one uses Gauss elimination which is usually applied to matrices, rather than systems of equations.

In [**Mak82**], Makanin reduced the decidability of equations over a free group to the decidability of finitely many systems of equations over a free monoid, in other words, he reduced the compatibility problem for a free group to the compatibility problem for generalised equations. In fact, Makanin essentially proved that the study of solutions of systems of equations over free groups reduces to the study of solutions of generalised equations in the following sense: every solution of the system of equations S factors trough one of the solutions of one of the generalised equations and, conversely, every solution of the generalised equation extends to a solution of S.

A crucial fact for this reduction is that the set of solutions of a given system of equations S over a free group, defines only finitely many different cancellation schemes (cancellation trees). By each of these cancellation trees, one can construct a generalised equation.

The goal of this chapter is to generalise this approach to systems of equations over a partially commutative group $\mathbb{G}$.

In Section 3.1, we give the definition of a generalised equation over a monoid $\mathbb{M}$. Then, along these lines, we define the constrained generalised equation over a monoid $\mathbb{M}$. Informally, a constrained generalised equation is simply a system of equations over a monoid with some constrains imposed onto its variables. In our case, the monoid we work with is either a trace monoid (alias for a partially commutative monoid) or a free monoid and the constrains that we impose on the variables are $\leftrightarrows$-commutation, see Section 2.5.

Our aim is to reduce the study of solutions of systems of equations over partially commutative groups to the study of solutions of constrained generalised equations over a free monoid. We do this reduction in two steps.

In Section 3.2, we show that to a system of equations over a partially commutative group one can associate a finite collection of (constrained) generalised equations over a partially commutative monoid. The family of solutions of the collection of generalised equations constructed describes all solutions of the initial system over a partially commutative group, see Lemma 3.16.

This reduction is performed using an analogue of the notion of a cancellation tree for free groups. Let $S(X) = 1$ be an equation over $\mathbb{G}$. Then for any solution $(g_1, \ldots, g_n) \in \mathbb{G}^n$ of S, the word $S(g_1, \ldots, g_n)$ represents a trivial element in $\mathbb{G}$. Thus, by van Kampen's lemma there exists a van Kampen diagram for this word. In the case of partially commutative groups van Kampen diagrams have a structure of a band complex, see [**CK07**]. We show in Proposition 2.7 that van Kampen diagrams over a partially commutative group can be taken to a "standard form". This standard form of van Kampen diagrams can be viewed as an analogue of the notion of a cancellation scheme. By Proposition 2.7, it follows that the set of solutions of a given system of equations S over a partially commutative group defines only finitely many van Kampen diagrams in standard form, i.e. finitely many different cancellation schemes. For each of these cancellation schemes one can construct a constrained generalised equation over the partially commutative monoid $\mathbb{T}$.

In Section 3.3 we show that for a given generalised equation over $\mathbb{T}$ one can associate a finite collection of (constrained) generalised equations over the free monoid $\mathbb{F}$. The family of solutions of the generalised equations from this collection describes all solutions of the initial generalised equation over $\mathbb{T}$, see Lemma 3.21.

This reduction relies on the ideas of Yu. Matiyasevich, see [**Mat97**] (see also [**DMM99**]) and V. Diekert and A. Muscholl, see [**DM06**]. Essentially, it states that there are finitely many ways to take the product of words (written in DM-normal form) in $\mathbb{T}$ to DM-normal form, see Proposition 3.18 and Corollary 3.19. We apply these results to reduce the study of the solutions of generalised equations over the trace monoid to the study of solutions of constrained generalised equations over a free monoid.

Finally, in Section 3.4 we give an example that follows the exposition of Chapter 3. We advise the reader unfamiliar with the terminology, to read the example of Section 3.4 simultaneously with the rest of Chapter 3.

We would like to mention that in [**DM06**] V. Diekert and A. Muscholl give a reduction of the compatibility problem of equations over a partially commutative group $\mathbb{G}$ to the decidability of equations over a free monoid with constraints. The reduction given in [**DM06**] provides a solution only to the compatibility problem of systems of equations over $\mathbb{G}$ and uses the theory of formal languages. In this chapter we employ the machinery of generalised equations in order to reduce the description of the set of solutions of a system over $\mathbb{G}$ to the same problem for constrained generalised equations over a free monoid and obtain a convenient setting for a further development of the process.

3.1. Definition of (constrained) generalised equations

Let $X = \{x_1, \ldots, x_n\}$ be a set of variables and let $\mathbb{G} = \mathbb{G}(\mathcal{A})$ be the partially commutative group generated by $\mathcal{A}$ and $\mathbb{G}[X] = \mathbb{G} * F(X)$.

Further by $\mathbb{M}$ we always mean either $\mathbb{F}(\mathcal{A}^{\pm 1})$ or $\mathbb{T}(\mathcal{A}^{\pm 1})$.

DEFINITION 3.1. A *combinatorial generalised equation* Υ over $\mathbb{M}$ (with coefficients from $\mathcal{A}^{\pm 1}$) consists of the following objects:

(1) A finite set of *bases* $\mathcal{BS} = \mathcal{BS}(\Upsilon)$. Every base is either a constant base or a variable base. Each constant base is associated with exactly one letter from $\mathcal{A}^{\pm 1}$. The set of variable bases $\mathcal{M}$ consists of $2n$ elements $\mathcal{M} = \{\mu_1, \ldots, \mu_{2n}\}$. The set $\mathcal{M}$ comes equipped with two functions: a function $\varepsilon : \mathcal{M} \to \{-1, 1\}$ and an involution without fixed points $\Delta : \mathcal{M} \to \mathcal{M}$ (i.e. Δ is a bijection such that Δ^2 is the identity on $\mathcal{M}$ and $\Delta(\mu) \neq \mu$ for all $\mu \in \mathcal{M}$). Bases μ and $\Delta(\mu)$ are called *dual bases*.

(2) A set of *boundaries* $\mathcal{BD} = \mathcal{BD}(\Upsilon)$. The set $\mathcal{BD}$ is a finite initial segment of the set of positive integers $\mathcal{BD} = \{1, 2, \ldots, \rho_\Upsilon + 1\}$.

(3) Two functions $\alpha : \mathcal{BS} \to \mathcal{BD}$ and $\beta : \mathcal{BS} \to \mathcal{BD}$. These functions satisfy the following conditions: $\alpha(b) < \beta(b)$ for every base $b \in \mathcal{BS}$; if b is a constant base then $\beta(b) = \alpha(b) + 1$.

(4) A finite set of *boundary connections* $\mathcal{BC} = \mathcal{BC}(\Upsilon)$. A $(\mu\text{-})$boundary connection is a triple (i, μ, j) where $i, j \in \mathcal{BD}$, $\mu \in \mathcal{M}$ such that $\alpha(\mu) < i < \beta(\mu)$, $\alpha(\Delta(\mu)) < j < \beta(\Delta(\mu))$ We assume that if $(i, \mu, j) \in \mathcal{BC}$ then $(j, \Delta(\mu), i) \in \mathcal{BC}$. This allows one to identify the boundary connections (i, μ, j) and $(j, \Delta(\mu), i)$.

Though, by the definition, a combinatorial generalised equation is a combinatorial object, it is not practical to work with combinatorial generalised equations describing its sets and functions. It is more convenient to encode all this information in its graphic representation. We refer the reader to Section 3.4 for the construction of a graphic representation of a generalised equation. All examples given in this paper use the graphic representation of generalised equations.

To a combinatorial generalised equation Υ over a monoid $\mathbb{M}$, one can associate a system of equations S_Υ in variables $h_1, \ldots, h_\rho$, $\rho = \rho_\Upsilon$ and coefficients from $\mathcal{A}^{\pm 1}$ (variables h_i are sometimes called *items*). The system of equations S_Υ consists of the following three types of equations.

(1) Each pair of dual variable bases $(\lambda, \Delta(\lambda))$ provides an equation over the monoid $\mathbb{M}$:

$$[h_{\alpha(\lambda)} h_{\alpha(\lambda)+1} \cdots h_{\beta(\lambda)-1}]^{\varepsilon(\lambda)} = [h_{\alpha(\Delta(\lambda))} h_{\alpha(\Delta(\lambda))+1} \cdots h_{\beta(\Delta(\lambda))-1}]^{\varepsilon(\Delta(\lambda))}.$$

These equations are called *basic equations*. In the case when $\beta(\lambda) = \alpha(\lambda)+1$ and $\beta(\Delta(\lambda)) = \alpha(\Delta(\lambda))+1$, i.e. the corresponding basic equation takes the form:

$$[h_{\alpha(\lambda)}]^{\varepsilon(\lambda)} = [h_{\alpha(\Delta(\lambda))}]^{\varepsilon(\Delta(\lambda))}.$$

(2) For each constant base b we write down a *coefficient equation* over $\mathbb{M}$:
$$h_{\alpha(b)} = a,$$
where $a \in \mathcal{A}^{\pm 1}$ is the constant associated to b.

(3) Every boundary connection (p, λ, q) gives rise to a *boundary equation* over $\mathbb{M}$, either
$$[h_{\alpha(\lambda)} h_{\alpha(\lambda)+1} \cdots h_{p-1}] = [h_{\alpha(\Delta(\lambda))} h_{\alpha(\Delta(\lambda))+1} \cdots h_{q-1}],$$
if $\varepsilon(\lambda) = \varepsilon(\Delta(\lambda))$, or
$$[h_{\alpha(\lambda)} h_{\alpha(\lambda)+1} \cdots h_{p-1}] = [h_q h_{q+1} \cdots h_{\beta(\Delta(\lambda))-1}]^{-1},$$
if $\varepsilon(\lambda) = -\varepsilon(\Delta(\lambda))$.

Conversely, given a system of equations $S(X, \mathcal{A}) = S$ over a monoid $\mathbb{M}$, one can construct a combinatorial generalised equation $\Upsilon(S)$ over $\mathbb{M}$.

Let $S = \{L_1 = R_1, \ldots, L_m = R_m\}$ be a system of equations over a monoid $\mathbb{M}$. Write S as follows:
$$l_{11} \cdots l_{1i_1} = r_{11} \cdots r_{1j_1}$$
$$\cdots$$
$$l_{m1} \cdots l_{mi_m} = r_{m1} \cdots r_{mj_m}$$

where $l_{ij}, r_{ij} \in X^{\pm 1} \cup \mathcal{A}^{\pm 1}$. The set of boundaries $\mathcal{BD}(\Upsilon(S))$ of the generalised equation $\Upsilon(S)$ is
$$\mathcal{BD}(\Upsilon(S)) = \left\{ 1, 2, \ldots, \sum_{k=1}^{m} (i_k + j_k) + 1 \right\}.$$

For all $k = 1, \ldots, \mathbb{m}$, we introduce a pair of dual variable bases $\mu_k, \Delta(\mu_k)$, so that
$$\alpha(\mu_k) = \sum_{n=1}^{k-1}(i_n + j_n) + 1, \quad \beta(\mu_k) = \alpha(\mu_k) + i_k, \quad \varepsilon(\mu_k) = 1;$$
$$\alpha(\Delta(\mu_k)) = \alpha(\mu_k) + i_k, \quad \beta(\Delta(\mu_k)) = \alpha(\Delta(\mu_k)) + j_k, \quad \varepsilon(\Delta(\mu_k)) = 1.$$

For any pair of distinct occurrences of a variable $x \in X$ as $l_{ij} = x^{\epsilon_{ij}}$, $l_{rs} = x^{\epsilon_{st}}$, $\epsilon_{ij}, \epsilon_{st} \in \{\pm 1\}$, where (i,j) precedes (s,t) in left-lexicographical order, we introduce a pair of dual bases $\mu_{x,q}, \Delta(\mu_{x,q})$, where $q = (i,j,s,t)$ so that
$$\alpha(\mu_{x,q}) = \sum_{n=1}^{i-1}(i_n + j_n) + j, \quad \beta(\mu_{x,q}) = \alpha(\mu_{x,q}) + 1, \quad \varepsilon(\mu_{x,q}) = \epsilon_{ij};$$
$$\alpha(\Delta(\mu_{x,q})) = \sum_{n=1}^{s-1}(i_n + j_n) + t, \quad \beta(\Delta(\mu_{x,q})) = \alpha(\mu_{x,q}) + 1, \quad \varepsilon(\Delta(\mu_{x,q})) = \epsilon_{st}.$$

Analogously, for any two occurrences of a variable $x \in X$ in S as $r_{ij} = x^{\epsilon_{ij}}$, $r_{st} = x^{\epsilon_{st}}$ or as $r_{ij} = x^{\epsilon_{ij}}$, $l_{st} = x^{\epsilon_{st}}$, we introduce the corresponding pair of dual bases.

For any occurrence of a constant $a \in \mathcal{A}$ in S as $l_{ij} = a^{\epsilon_{ij}}$ we introduce a constant base ν_a so that
$$\alpha(\nu_a) = \sum_{n=1}^{i-1}(i_n + j_n) + j, \quad \beta(\nu_a) = \alpha(\nu_a) + 1.$$

Similarly, for any occurrence of a constant $a \in \mathcal{A}$ as $r_{st} = a^{\epsilon_{st}}$, we introduce a constant base ν_a so that

$$\alpha(\nu_a) = \sum_{n=1}^{s-1}(i_n + j_n) + i_s + t, \quad \beta(\nu_a) = \alpha(\nu_a) + 1.$$

The set of boundary connections $\mathcal{BC}$ is empty.

DEFINITION 3.2. Introduce an equivalence relation on the set of all combinatorial generalised equations over $\mathbb{M}$ as follows. Two generalised equations Υ and Υ' are *equivalent*, in which case we write $\Upsilon \approx \Upsilon'$, if and only if the corresponding systems of equations S_Υ and $S_{\Upsilon'}$ are equivalent (recall that two systems are called equivalent if their sets of solutions are isomorphic).

LEMMA 3.3. *There is a one-to-one correspondence between the set of $\approx$-equivalence classes of combinatorial generalised equations over $\mathbb{M}$ and the set of equivalence classes of systems of equations over $\mathbb{M}$. Furthermore, this correspondence is given effectively.*

PROOF. Define the correspondence between the set of $\approx$-equivalence classes of combinatorial generalised equations over $\mathbb{M}$ and the set of equivalence classes of systems of equations over $\mathbb{M}$ as follows. To a combinatorial generalised equation Υ over $\mathbb{M}$ we assign the system of equations S_Υ associated to Υ and to a system of equations S over $\mathbb{M}$ we assign the combinatorial generalised equation $\Upsilon(S)$ associated to it. We now prove that this correspondence is well-defined.

Let Υ and Υ' be two $\approx$-equivalent generalised equations and let S_Υ and $S_{\Upsilon'}$ be the corresponding associated systems of equations over $\mathbb{M}$. Then, by definition of the equivalence relation '$\approx$', one has that S_Υ and $S_{\Upsilon'}$ are equivalent.

Conversely, let S and S' be two equivalent systems of equations over $\mathbb{M}$ and let $\Upsilon(S)$ and $\Upsilon(S')$ be the corresponding associated combinatorial generalised equations over $\mathbb{M}$. By definition, $\Upsilon(S)$ and $\Upsilon(S')$ are equivalent if and only if the systems $S_{\Upsilon(S)}$ and $S_{\Upsilon(S')}$ are. By construction, it is easy to check, that the system of equations $S_{\Upsilon(S)}$ associated to $\Upsilon(S)$ is equivalent to S and the system $S_{\Upsilon(S')}$ is equivalent to S', thus, by transitivity, $S_{\Upsilon(S)}$ is equivalent to $S_{\Upsilon(S')}$. □

Abusing the language, we call the system S_Υ associated to a generalised equation Υ *generalised equation* over $\mathbb{M}$, and, abusing the notation, we further denote it by the same symbol Υ.

DEFINITION 3.4. A *constrained generalised equation* Ω over $\mathbb{M}$ is a pair $\langle \Upsilon, \Re_\Upsilon \rangle$, where Υ is a generalised equation and $\Re_\Upsilon$ is a symmetric binary relation on the set of variables $h_1, \ldots, h_\rho$ of the generalised equation Υ that satisfies the following condition.

($\star$) Let $\Re_\Upsilon(h_i) = \{h_j \mid \Re_\Upsilon(h_i, h_j)\}$. If in Υ there is an equation of the form

$$h_{i_1}^{\epsilon_{i_1}} \ldots h_{i_k}^{\epsilon_{i_k}} = h_{j_1}^{\epsilon_{j_1}} \ldots h_{j_l}^{\epsilon_{j_l}}, \quad \epsilon_{i_n}, \epsilon_{j_t}, \in \{1, -1\}, \quad n = 1, \ldots, k, \, t = 1, \ldots, l$$

and there exists h_m such that $\Re_\Upsilon(h_{i_n}, h_m)$ for all $n = 1, \ldots, k$, then $\Re_\Upsilon(h_{j_t}, h_m)$ for all $t = 1, \ldots, l$.

DEFINITION 3.5. Let $\Upsilon(h) = \{L_1(h) = R_1(h), \ldots, L_s(h) = R_s(h)\}$ be a generalised equation over $\mathbb{T}$ in variables $h = (h_1, \ldots, h_\rho)$ with coefficients from $\mathbb{G}$. A tuple $H = (H_1, \ldots, H_\rho)$ of non-empty words from $\mathbb{G}$ in the normal form (see Section 2.5) is a *solution* of Υ if:

(1) all words $L_i(H), R_i(H)$ are geodesic (treated as elements of $\mathbb{G}$);
(2) $L_i(H) = R_i(H)$ in the monoid $\mathbb{T}$ for all $i = 1, \ldots, s$.

DEFINITION 3.6. Let $\Upsilon(h) = \{L_1(h) = R_1(h), \ldots, L_s(h) = R_s(h)\}$ be a generalised equation over $\mathbb{F}$ in variables $h = (h_1, \ldots, h_\rho)$ with coefficients from $\mathbb{G}$. A tuple $H = (H_1, \ldots, H_\rho)$ of non-empty geodesic words from $\mathbb{G}$ is a *solution* of Υ if:
(1) all words $L_i(H), R_i(H)$ are geodesic (treated as elements of $\mathbb{G}$);
(2) $L_i(H) = R_i(H)$ in $\mathbb{F}$ for all $i = 1, \ldots, s$.

The notation (Υ, H) means that H is a solution of the generalised equation Υ.

DEFINITION 3.7. Let $\Omega = \langle \Upsilon, \Re_\Upsilon \rangle$ be a constrained generalised equation over $\mathbb{M}$ in variables $h = (h_1, \ldots, h_\rho)$ with coefficients from $\mathbb{G}$. A tuple $H = (H_1, \ldots, H_\rho)$ of non-empty geodesic words in $\mathbb{G}$ is a *solution* of Ω if H is a solution of the generalised equation Υ and $H_i \leftrightarrows H_j$ if $\Re_\Upsilon(h_i, h_j)$.

The *length* of a solution H is defined to be

$$|H| = \sum_{i=1}^{\rho} |H_i|.$$

The notation (Ω, H) means that H is a solution of the constrained generalised equation Ω.

The term "constrained generalised equation over $\mathbb{M}$" can be misleading. We would like to stress that solutions of a constrained generalised equations are *not* just solutions of the system of equations (associated to the generalised equation) over $\mathbb{M}$. They are tuples of non-empty geodesic words from $\mathbb{G}$ such that substituting these words in to the equations of a generalised equation, one gets equalities in $\mathbb{M}$.

NOTA BENE. Further we abuse the terminology and call Ω simply "generalised equation". However, we always use the symbol Ω for constrained generalised equations and Υ for generalised equations.

We now introduce a number of notions that we use throughout the text. Let Ω be a generalised equation.

DEFINITION 3.8 (Glossary of terms). A boundary i *intersects* the base μ if $\alpha(\mu) < i < \beta(\mu)$. A boundary i *touches* the base μ if $i = \alpha(\mu)$ or $i = \beta(\mu)$. A boundary is said to be *open* if it intersects at least one base, otherwise it is called *closed*. We say that a boundary i is *tied* in a base μ (or is μ-*tied*) if there exists a boundary connection (p, μ, q) such that $i = p$ or $i = q$. A boundary is *free* if it does not touch any base and it is not tied by a boundary connection.

An item h_i *belongs* to a base μ or, equivalently, μ *contains* h_i, if $\alpha(\mu) \leq i \leq \beta(\mu) - 1$ (in this case we sometimes write $h_i \in \mu$). An item h_i is called a *constant* item if it belongs to a constant base and h_i is called a *free* item if it does not belong to any base. By $\gamma(h_i) = \gamma_i$ we denote the number of bases which contain h_i, in this case we also say that h_i is *covered* γ_i times. An item h_i is called *linear* if $\gamma_i = 1$ and is called *quadratic* if $\gamma_i = 2$.

Let $\mu, \Delta(\mu)$ be a pair of dual bases such that $\alpha(\mu) = \alpha(\Delta(\mu))$ and $\beta(\mu) = \beta(\Delta(\mu))$ in this case we say that bases μ and $\Delta(\mu)$ form *a pair of matched bases*. A base λ is *contained* in a base μ if $\alpha(\mu) \leq \alpha(\lambda) < \beta(\lambda) \leq \beta(\mu)$. We say that two bases μ and ν *intersect* or *overlap*, if $[\alpha(\mu), \beta(\mu)] \cap [\alpha(\nu), \beta(\nu)] \neq \emptyset$. A base μ is called *linear* if there exists an item $h_i \in \mu$ so that h_i is linear.

3.1. DEFINITION OF (CONSTRAINED) GENERALISED EQUATIONS

A set of consecutive items $[i,j] = \{h_i, \ldots, h_{j-1}\}$ is called a *section*. A section is said to be *closed* if the boundaries i and j are closed and all the boundaries between them are open. If μ is a base then by $\sigma(\mu)$ we denote the section $[\alpha(\mu), \beta(\mu)]$ and by $h(\mu)$ we denote the product of items $h_{\alpha(\mu)} \ldots h_{\beta(\mu)-1}$. In general for a section $[i,j]$ by $h[i,j]$ we denote the product $h_i \ldots h_{j-1}$. A base μ *belongs* to a section $[i,j]$ if $i \leq \alpha(\mu) < \beta(\mu) \leq j$. Similarly an item h_k *belongs* to a section $[i,j]$ if $i \leq k < j$. In these cases we write $\mu \in [i,j]$ or $h_k \in [i,j]$.

Let $H = (H_1, \ldots, H_\rho)$ be a solution of a generalised equation Ω in variables $h = \{h_1, \ldots, h_\rho\}$. We use the following notation. For any word $W(h)$ in $\mathbb{G}[h]$ set $W(H) = H(W(h))$. In particular, for any base μ (section $\sigma = [i,j]$) of Ω, we have $H(\mu) = H(h(\mu)) = H_{\alpha(\mu)} \cdots H_{\beta(\mu)-1}$ ($H[i,j] = H(\sigma) = H(h(\sigma)) = H_i \cdots H_{j-1}$, respectively).

We now formulate some necessary conditions for a generalised equation to have a solution.

DEFINITION 3.9. A generalised equation $\Omega = \langle \Upsilon, \Re_\Upsilon \rangle$ is called *formally consistent* if it satisfies the following conditions.

(1) If $\varepsilon(\mu) = -\varepsilon(\Delta(\mu))$, then the bases μ and $\Delta(\mu)$ do not intersect, i.e. $[\alpha(\mu), \beta(\mu)] \cap [\alpha(\Delta(\mu)), \beta(\Delta(\mu))] = \emptyset$.
(2) Given two boundary connections (p, λ, q) and (p_1, λ, q_1), if $p \leq p_1$, then $q \leq q_1$ in the case when $\varepsilon(\lambda)\varepsilon(\Delta(\lambda)) = 1$, and $q \geq q_1$ in the case when $\varepsilon(\lambda)\varepsilon(\Delta(\lambda)) = -1$. In particular, if $p = p_1$ then $q = q_1$.
(3) Let μ be a base such that $\alpha(\mu) = \alpha(\Delta(\mu))$, in other words, let μ and $\Delta(\mu)$ be a pair of matched bases. If (p, μ, q) is a μ-boundary connection then $p = q$.
(4) A variable cannot occur in two distinct coefficient equations, i.e., any two constant bases with the same left end-point are labelled by the same letter from $\mathcal{A}^{\pm 1}$.
(5) If h_i is a variable from some coefficient equation and $(i, \mu, q_1), (i+1, \mu, q_2)$ are boundary connections, then $|q_1 - q_2| = 1$.
(6) If $\Re_\Upsilon(h_i, h_j)$ then $i \neq j$.

LEMMA 3.10.

(1) *If a generalised equation Ω over the monoid $\mathbb{M}$ has a solution, then Ω is formally consistent;*
(2) *There is an algorithm to check whether or not a given generalised equation is formally consistent.*

PROOF. We show that condition (1) of Definition 3.9 holds for the generalised equation Ω in the case $\mathbb{M} = \mathbb{T}$. Assume the contrary, i.e. $\varepsilon(\mu) = -\varepsilon(\Delta(\mu))$ and the bases μ and $\Delta(\mu)$ intersect. Let H be a solution of Ω. Without loss of generality we may assume that $\varepsilon(\mu) = 1$ and $\alpha(\mu) \leq \alpha(\Delta(\mu))$. Then Ω has the following basic equation:
$$h_{\alpha(\mu)} \cdots h_{\beta(\mu)-1} = \left(h_{\alpha(\Delta(\mu))} \cdots h_{\beta(\Delta(\mu))-1}\right)^{-1}.$$
Since the bases μ and $\Delta(\mu)$ intersect, the words $H[\alpha(\Delta(\mu)), \beta(\mu)]$ and $H[\alpha(\Delta(\mu)), \beta(\mu)]^{-1}$ right-divide the word $H(\mu)$. This derives a contradiction with the fact that $H(\mu)$ is a geodesic word in $\mathbb{G}$, see [**EKR05**].

Proof follows by straightforward verification of the conditions in Definition 3.9. □

REMARK 3.11. We further consider only formally consistent generalised equations.

3.2. Reduction to generalised equations: from partially commutative groups to monoids

In this section we show that to a given finite system of equations $S(X,\mathcal{A}) = 1$ over a partially commutative group $\mathbb{G}$ one can associate a finite collection of (constrained) generalised equations $\mathcal{GE}'(S)$ over $\mathbb{T}$ with coefficients from $\mathcal{A}^{\pm 1}$. The family of solutions of the generalised equations from $\mathcal{GE}'(S)$ describes all solutions of the system $S(X,\mathcal{A}) = 1$, see Lemma 3.16.

3.2.1. $\mathbb{G}$-partition tables.
Write the system $\{S(X,\mathcal{A}) = 1\} = \{S_1 = 1, \ldots, S_{\mathrm{m}} = 1\}$ in the form

(3.1)
$$\begin{aligned} r_{11}r_{12}\ldots r_{1l_1} &= 1, \\ r_{21}r_{22}\ldots r_{2l_2} &= 1, \\ &\ldots \\ r_{\mathrm{m}1}r_{\mathrm{m}2}\ldots r_{\mathrm{m}l_\mathrm{m}} &= 1, \end{aligned}$$

where r_{ij} are letters of the alphabet $X^{\pm 1} \cup \mathcal{A}^{\pm 1}$.

We aim to define a combinatorial object called a $\mathbb{G}$-partition table, that encodes a particular type of cancellation that happens when one substitutes a solution $W(\mathcal{A}) \in \mathbb{G}^n$ into $S(X,\mathcal{A}) = 1$ and then reduces the words in $S(W(\mathcal{A}),\mathcal{A})$ to the empty word.

Informally, Proposition 2.7 describes all possible cancellation schemes for the set of all solutions of the system $S(X,\mathcal{A})$ in the following way: the cancellation scheme corresponding to a particular solution, can be obtained from the one described in Proposition 2.7 by setting some of the words w_i^j's (and the corresponding bands) to be trivial. Therefore, every $\mathbb{G}$-partition table (to be defined below) corresponds to one of the cancellation schemes obtained from the general one by setting some of the words w_i^j's to be trivial. Every non-trivial word w_i^j corresponds to a variable z_k and the word w_j^i to the variable z_k^{-1}. If a variable x that occurs in the system $S(X,A) = 1$ is subdivided into a product of some words w_i^j's, i.e. the variable x is a word in the w_i^j's, then the word V_{ij} from the definition of a partition table is this word in the corresponding z_k's. If the bands corresponding to the words w_i^j and w_k^l cross, then the corresponding variables z_r and z_s commute in the group $\mathbb{H}$.

The definition of a $\mathbb{G}$-partition table is rather technical, we refer the reader to Section 3.4 for an example.

A pair $\mathcal{T}$ (a finite set of geodesic words from $\mathbb{G} * F(Z)$, a $\mathbb{G}$-partially commutative group), $\mathcal{T} = (V, \mathbb{H})$ of the form:

$$V = \{V_{ij}(z_1, \ldots, z_p)\} \subset \mathbb{G}[Z] = \mathbb{G} * F(Z) \ (1 \leq i \leq m, 1 \leq j \leq l_i), \ \mathbb{H} = \mathbb{G}(\mathcal{A} \cup Z),$$

is called a $\mathbb{G}$-*partition table* of the system $S(X,\mathcal{A})$ if the following conditions are satisfied:

(1) Every element $z \in Z \cup Z^{-1}$ occurs in the words V_{ij} only once;
(2) The equality $V_{i1}V_{i2}\cdots V_{il_i} = 1, 1 \leq i \leq \mathrm{m}$, holds in $\mathbb{H}$;
(3) $|V_{ij}| \leq l_i - 1$;
(4) if $r_{ij} = a \in \mathcal{A}^{\pm 1}$, then $|V_{ij}| = 1$.

Here the designated copy of $\mathbb{G}$ in $\mathbb{H}$ is the natural one and is generated by $\mathcal{A}$.

Since $|V_{ij}| \leq l_i - 1$ then at most $\sum_{i=1}^{m}(l_i - 1)l_i$ different letters z_i can occur in a partition table of $S(X, \mathcal{A}) = 1$. Therefore we always assume that $p \leq \sum_{i=1}^{m}(l_i - 1)l_i$.

REMARK 3.12. In the case that $\mathbb{G} = F(\mathcal{A})$ and $\mathbb{H} = F(Z)$ are free groups, the notion of a $\mathbb{G}$-partition table coincides with the notion of a partition table in the sense of Makanin, see [**Mak82**].

LEMMA 3.13. *Let $S(X, \mathcal{A}) = 1$ be a finite system of equations over $\mathbb{G}$. Then*
 (1) *the set $\mathcal{PT}(S)$ of all $\mathbb{G}$-partition tables of $S(X, \mathcal{A}) = 1$ is finite, and its cardinality is bounded by a number which depends only on the system $S(X, \mathcal{A})$;*
 (2) *one can effectively enumerate the set $\mathcal{PT}(S)$.*

PROOF. Since the words V_{ij} have bounded length, one can effectively enumerate the finite set of all collections of words $\{V_{ij}\}$ in $F(Z)$ which satisfy the conditions (1), (3), (4) above. Now for each such collection $\{V_{ij}\}$, one can effectively check whether the equalities $V_{i1}V_{i2}\cdots V_{il_i} = 1, 1 \leq i \leq m$ hold in one of the finitely many (since $|Z| < \infty$) partially commutative groups $\mathbb{H}$ or not. This allows one to list effectively all partition tables for $S(X, \mathcal{A}) = 1$. □

3.2.2. Associating a generalised equation over $\mathbb{T}$ to a $\mathbb{G}$-partition table. By a partition table $\mathcal{T} = (\{V_{ij}\}, \mathbb{H})$ we construct a generalised equation $\Omega'_{\mathcal{T}} = \langle \Upsilon'_{\mathcal{T}}, \Re_{\Upsilon'_{\mathcal{T}}} \rangle$ in the following way. We refer the reader to Section 3.4 for an example.

Consider the following word $\mathcal{V}$ in $\mathbb{F}(Z^{\pm 1})$:
$$\mathcal{V} \doteq V_{11}V_{12}\cdots V_{1l_1}\cdots V_{m1}V_{m2}\cdots V_{ml_m} = y_1\cdots y_{\rho'},$$
where $\mathbb{F}(Z^{\pm 1})$ is the free monoid on the alphabet $Z^{\pm 1}$; $y_i \in Z^{\pm 1}$ and $\rho' = |\mathcal{V}|$ is the length of $\mathcal{V}$. Then the generalised equation $\Omega'_{\mathcal{T}} = \Omega'_{\mathcal{T}}(h')$ has $\rho' + 1$ boundaries and ρ' variables $h'_1, \ldots, h'_{\rho'}$ which are denoted by $h' = (h'_1, \ldots, h'_{\rho'})$.

Now we define the bases of $\Omega'_{\mathcal{T}}$ and the functions $\alpha, \beta, \varepsilon$.

Let $z \in Z$. For the (unique) pair of distinct occurrences of z in $\mathcal{V}$:
$$y_i = z^{\epsilon_i}, \quad y_j = z^{\epsilon_j} \quad \epsilon_i, \epsilon_j \in \{1, -1\}, \quad i < j,$$
we introduce a pair of dual variable bases $\mu_{z,i}, \mu_{z,j}$ such that $\Delta(\mu_{z,i}) = \mu_{z,j}$. Put
$$\alpha(\mu_{z,i}) = i, \quad \beta(\mu_{z,i}) = i+1, \quad \varepsilon(\mu_{z,i}) = \epsilon_i,$$
$$\alpha(\mu_{z,j}) = j, \quad \beta(\mu_{z,j}) = j+1, \quad \varepsilon(\mu_{z,j}) = \epsilon_j.$$

The basic equation that corresponds to this pair of dual bases is $h'^{\epsilon_i}_i \doteq h'^{\epsilon_j}_j$.

Let $x \in X$. For any two distinct occurrences of x in $S(X, \mathcal{A}) = 1$:
$$r_{i,j} = x^{\epsilon_{ij}}, \quad r_{s,t} = x^{\epsilon_{st}} \quad (\epsilon_{ij}, \epsilon_{st} \in \{1, -1\})$$
so that (i, j) precedes (s, t) in left-lexicographical order, we introduce a pair of dual bases $\mu_{x,q}$ and $\Delta(\mu_{x,q})$, $q = (i, j, s, t)$. Now suppose that V_{ij} and V_{st} occur in the word $\mathcal{V}$ as subwords
$$V_{ij} = y_{c_1}\cdots y_{d_1}, \quad V_{st} = y_{c_2}\cdots y_{d_2} \text{ correspondingly.}$$

Then we put
$$\alpha(\mu_{x,q}) = c_1, \quad \beta(\mu_{x,q}) = d_1 + 1, \quad \varepsilon(\mu_{x,q}) = \epsilon_{ij},$$
$$\alpha(\Delta(\mu_{x,q})) = c_2, \quad \beta(\Delta(\mu_{x,q})) = d_2 + 1, \quad \varepsilon(\Delta(\mu_{x,q})) = \epsilon_{st}.$$

The basic equation over $\mathbb{T}$ which corresponds to this pair of dual bases can be written in the form
$$\left(h'_{\alpha(\mu_{x,q})} \cdots h'_{\beta(\mu_{x,q})-1}\right)^{\epsilon_{ij}} = \left(h'_{\alpha(\Delta(\mu_{x,q}))} \cdots h'_{\beta(\Delta(\mu_{x,q}))-1}\right)^{\epsilon_{st}}.$$

Let $r_{ij} = a \in \mathcal{A}^{\pm 1}$. In this case we introduce a constant base μ_{ij} with the label a. If V_{ij} occurs in $\mathcal{V}$ as $V_{ij} = y_c$, then we put
$$\alpha(\mu_{ij}) = c, \quad \beta(\mu_{ij}) = c + 1.$$

The corresponding coefficient equation is $h'_c = a$. The set of boundary connections of the generalised equation $\Upsilon'_\mathcal{T}$ is empty. This defines the generalised equation $\Upsilon'_\mathcal{T}$.

We define the binary relation $\Re_{\Upsilon'_\mathcal{T}} \subseteq h' \times h'$ to be the minimal subset of $h' \times h'$ which contains the pairs (h'_i, h'_j) such that $[y_i, y_j] = 1$ in $\mathbb{H}$ and $y_i \neq y_j$, is symmetric and satisfies condition $(\star)$ of Definition 3.4. This defines the constrained generalised equation $\Omega'_\mathcal{T} = \langle \Upsilon'_\mathcal{T}, \Re_{\Upsilon'_\mathcal{T}} \rangle$.

Put
$$\mathcal{GE}'(S) = \{\Omega'_\mathcal{T} \mid \mathcal{T} \text{ is a } \mathbb{G}\text{-partition table for } S(X, \mathcal{A}) = 1\}.$$

Then $\mathcal{GE}'(S)$ is a finite collection of generalised equations over $\mathbb{T}$ which can be effectively constructed for a given system of equations $S(X, \mathcal{A}) = 1$ over $\mathbb{G}$.

REMARK 3.14. As in Remark 3.12, if $\mathbb{G} = F(\mathcal{A})$ and $\mathbb{H} = F(Z)$ are free groups and the set $\Re_{\Upsilon'_\mathcal{T}}$ is empty, then the basic equations are considered over a free monoid and the collection of generalised equations $\mathcal{GE}'$ coincides with the one used by Makanin, see [**Mak82**].

3.2.3. Coordinate groups of systems of equations and coordinate groups of generalised equations.

DEFINITION 3.15. For a generalised equation Υ in variables h over a monoid $\mathbb{M}$ we can consider the same system of equations over the partially commutative group (not in the monoid). We denote this system by Υ^*. In other words, if
$$\Upsilon = \{L_1(h) = R_1(h), \ldots, L_s(h) = R_s(h)\}$$
is an arbitrary system of equations over $\mathbb{M}$ with coefficients from $\mathcal{A}^{\pm 1}$, then by Υ^* we denote the system of equations
$$\Upsilon^* = \{L_1(h)R_1(h)^{-1} = 1, \ldots, L_s(h)R_s(h)^{-1} = 1\}$$
over the group $\mathbb{G}$.

Similarly, for a given constrained generalised equation $\Omega = \langle \Upsilon, \Re_\Upsilon \rangle$ in variables h over a monoid $\mathbb{M}$ we can consider the system of equations $\Omega^* = \Upsilon^* \cup \{[h_i, h_j] \mid \Re_\Upsilon(h_i, h_j)\}$ over the partially commutative group (not in the monoid). Let
$$\mathbb{G}_{R(\Omega^*)} = \mathbb{G} * F(h) \big/ R(\Upsilon^* \cup \{[h_i, h_j] \mid \Re_\Upsilon(h_i, h_j)\}),$$
where $F(h)$ is the free group with basis h. We call $\mathbb{G}_{R(\Omega^*)}$ the *coordinate group* of the constrained generalised equation Ω.

3.2. FROM PARTIALLY COMMUTATIVE GROUPS TO MONOIDS

Note that the definition of the coordinate group of a constrained generalised equation over $\mathbb{M}$ is independent of the monoid $\mathbb{M}$.

Obviously, each solution H of Ω over $\mathbb{M}$ gives rise to a solution of Ω^* in the partially commutative group $\mathbb{G}$. The converse does not hold in general even in the case when $\mathbb{M} = \mathbb{T}$: it may happen that H is a solution of Ω^* in $\mathbb{G}$ but not in $\mathbb{T}$, i.e. some equalities $L_i(H) = R_i(H)$ hold only after a reduction in $\mathbb{G}$ (and, certainly, the equation $[h_i, h_j]$ does not imply the $\leftrightarrows$-commutation for the solution).

Now we explain the relation between the coordinate groups $\mathbb{G}_{R(\Omega'_{\mathcal{T}}*)}$ and $\mathbb{G}_{R(S(X,\mathcal{A}))}$.

For a letter x in X we choose an arbitrary occurrence of x in $S(X, \mathcal{A}) = 1$ as
$$r_{ij} = x^{\epsilon_{ij}}.$$
Let $\mu = \mu_{x,q}$, $q = (i, j, s, t)$ be a base that corresponds to this occurrence of x. Then V_{ij} occurs in $\mathcal{V}$ as the subword
$$V_{ij} = y_{\alpha(\mu)} \cdots y_{\beta(\mu)-1}.$$
Notice that the word V_{ij} does not depend on the choice of the base $\mu_{x,q}$ corresponding to the occurrence r_{ij}. Define a word $P_x(h') \in \mathbb{G}[h']$ (where $h' = \{h'_1, \ldots, h'_{\rho'}\}$) as follows
$$P_x(h', \mathcal{A}) = \left(h'_{\alpha(\mu)} \cdots h'_{\beta(\mu)-1}\right)^{\epsilon_{ij}},$$
and put
$$P(h') = (P_{x_1}, \ldots, P_{x_n}).$$
The word $P_x(h')$ depends on the choice of occurrence $r_{ij} = x^{\epsilon_{ij}}$ in $\mathcal{V}$.

It follows from the construction above that the map $X \to \mathbb{G}[h']$ defined by $x \mapsto P_x(h', \mathcal{A})$ gives rise to a $\mathbb{G}$-homomorphism

(3.2) $$\pi : \mathbb{G}_{R(S)} \to \mathbb{G}_{R(\Omega'_{\mathcal{T}}*)}.$$

Indeed, if $f(X) \in R(S)$ then $\pi(f(X)) = f(P(h))$. It follows from condition (2) of the definition of partition table that $f(P(h)) = 1$ in $\mathbb{G}_{R(\Omega'_{\mathcal{T}}*)}$, thus $f(P(h)) \in R(\Omega'_{\mathcal{T}}*)$. Therefore $R(f(S)) \subseteq R(\Omega'_{\mathcal{T}}*)$ and π is a homomorphism.

Observe that the image $\pi(x)$ in $\mathbb{G}_{R(\Omega'_{\mathcal{T}}*)}$ does not depend on a particular choice of the occurrence of x in $S(X, \mathcal{A})$ (the basic equations of $\Omega'_{\mathcal{T}}$ make these images equal). Hence π depends only on $\Omega'_{\mathcal{T}}$. Thus, every solution H of a generalised equation gives rise to a solution U of S so that $\pi_U = \pi\pi_H$.

3.2.4. Solutions of systems of equations over $\mathbb{G}$ and solutions of generalised equations over $\mathbb{T}$. Our goal in this section is to prove that every solution of the system of equations $S(X, \mathcal{A})$ over $\mathbb{G}$ factors through one of the solutions of one of the finitely many generalised equations from $\mathcal{GE}'(S)$, i.e. for a solution W of S there exists a generalised equation $\Omega'_{\mathcal{T}} \in \mathcal{GE}'(S)$ and a solution H of $\Omega'_{\mathcal{T}}$ so that $\pi_W = \pi\pi_H$, where π is defined in (3.2). In order to do so, we need an analogue of the notion of a cancellation scheme for free groups. The results of this section rely on the techniques developed in Section 3 of [**CK07**].

Let $W(\mathcal{A})$ be a solution of $S(X, \mathcal{A}) = 1$ in $\mathbb{G}$. If in the system (3.1) we make the substitution $\sigma : X \to W(\mathcal{A})$, then
$$(r_{i1}r_{i2} \ldots r_{il_i})^\sigma = r^\sigma_{i1} r^\sigma_{i2} \ldots r^\sigma_{il_i} = 1$$
in $\mathbb{G}$ for every $i = 1, \ldots, \mathsf{m}$.

Since every product $R_i^\sigma = r_{i1}^\sigma r_{i2}^\sigma \ldots r_{il_i}^\sigma$ is trivial, we can construct a van Kampen diagram $\mathcal{D}_{R_i^\sigma}$ for R_i^σ. Denote by $\tilde{z}_{i,1}, \ldots, \tilde{z}_{i,p_i}$ the subwords w_j^k, $1 \leq j < k \leq l_i$ of r_{ij}^σ, where w_j^k are defined as in Proposition 2.7. As, by Proposition 2.7, $w_j^k = w_k^{j-1}$, so every r_{ij} can be written as a freely reduced word $V_{ij}(Z_i)$ in variables $Z_i = \{z_{i,1}, \ldots, z_{i,p_i}\}$ so that if we set $z_{i,j}^\sigma = \tilde{z}_{i,j}$, then we have that the following equality holds in $\mathbb{T}$:

$$r_{ij}^\sigma = V_{ij}(\tilde{z}_{i,1}, \ldots, \tilde{z}_{i,p_i}).$$

Observe that if $r_{ij} = a \in \mathcal{A}^{\pm 1}$ then $r_{ij}^\sigma = a$ and we have $|V_{ij}| = 1$. By Proposition 2.7, r_{ij}^σ is a product of at most $l_i - 1$ words w_j^k, thus we have that $|V_{ij}| \leq l_i - 1$. Denote by $Z = \bigcup_{i=1}^m Z_i = \{z_1, \ldots, z_p\}$. Take the partially commutative group $\mathbb{H} = \mathbb{G}(\mathcal{A} \cup Z)$ whose underlying commutation graph is defined as follows:

- two elements a_i, a_j in $\mathcal{A}$ commute whenever they commute in $\mathbb{G}$;
- an element $a \in \mathcal{A}$ commutes with z_i whenever $\tilde{z}_i \leftrightarrows a$;
- two elements $z_i, z_j \in Z$ commute whenever $\tilde{z}_i \leftrightarrows \tilde{z}_j$.

By construction, the set $\{V_{ij}\}$ along with the group $\mathbb{H}$ is a partition table $\mathcal{T}_{W(\mathcal{A})}$ for the system $S(X, \mathcal{A}) = 1$ and the solution $W(\mathcal{A})$. Obviously,

$$H' = (\tilde{z}_1, \ldots, \tilde{z}_p)$$

is the solution of the generalised equation $\Omega_{\mathcal{T}_{W(\mathcal{A})}}$ induced by $W(\mathcal{A})$. From the construction of the map $P(h')$ we deduce that $W(\mathcal{A}) = P(H')$.

The following lemma shows that to describe the set of solutions of a system of equations over $\mathbb{G}$ is equivalent to describe the set of solutions of constrained generalised equations over the trace monoid $\mathbb{T}$. This lemma can be viewed as the first "divide" step of the process.

LEMMA 3.16. *For a given system of equations $S(X, \mathcal{A}) = 1$ over $\mathbb{G}$, one can effectively construct a finite set*

$$\mathcal{GE}'(S) = \{\Omega_{\mathcal{T}}' \mid \mathcal{T} \text{ is a } \mathbb{G}\text{-partition table for } S(X, \mathcal{A}) = 1\}$$

of generalised equations over $\mathbb{T}$ such that

(1) *if the set $\mathcal{GE}'(S)$ is empty, then $S(X, \mathcal{A}) = 1$ has no solutions in $\mathbb{G}$;*
(2) *for each $\Omega'(h') \in \mathcal{GE}'(S)$ and for each $x \in X$ one can effectively find a word $P_x(h', \mathcal{A}) \in \mathbb{G}[h']$ of length at most $|h'|$ such that the map $x \mapsto P_x(h', \mathcal{A})$ gives rise to a $\mathbb{G}$-homomorphism $\pi_{\Omega'} : \mathbb{G}_{R(S)} \to \mathbb{G}_{R(\Omega'^*)}$ (in particular, for every solution H' of the generalised equation Ω' one has that $P(H')$ is a solution of the system $S(X, \mathcal{A})$, where $P(h') = (P_{x_1}, \ldots, P_{x_n})$);*
(3) *for any solution $W(\mathcal{A}) \in \mathbb{G}^n$ of the system $S(X, \mathcal{A}) = 1$ there exists $\Omega'(h') \in \mathcal{GE}'(S)$ and a solution H' of $\Omega'(h')$ such that $W(\mathcal{A}) = P(H')$, where $P(h') = (P_{x_1}, \ldots, P_{x_n})$, and this equality holds in the partially commutative monoid $\mathbb{T}(\mathcal{A}^{\pm 1})$;*

COROLLARY 3.17. *In the notation of Lemma 3.16 for any solution $W(\mathcal{A}) \in \mathbb{G}^n = \mathbb{G}(\mathcal{A})^n$ of the system $S(X, \mathcal{A}) = 1$ there exist a generalised equation $\Omega'(h') \in$*

$\mathcal{GE}'(S)$ and a solution H' of $\Omega'(h')$ such that the following diagram commutes

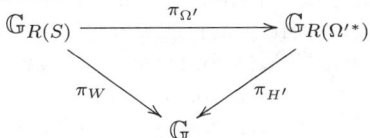

Conversely, for every generalised equation $\Omega'(h') \in \mathcal{GE}'(S)$ and a solution H' of $\Omega'(h')$ there exists a solution $W(\mathcal{A}) \in \mathbb{G}^n = \mathbb{G}(\mathcal{A})^n$ of the system $S(X, \mathcal{A}) = 1$ such that the above diagram commutes.

3.3. Reduction to generalised equations: from partially commutative monoids to free monoids

In this section we show that to a given generalised equation Ω' over $\mathbb{T}$ one can associate a finite collection of (constrained) generalised equations $\mathcal{GE}(\Omega')$ over $\mathbb{F}$ with coefficients from $\mathcal{A}^{\pm 1}$. The family of solutions of the generalised equations from $\mathcal{GE}(\Omega')$ describes all solutions of the generalised equation Ω', see Lemma 3.21.

We shall make use of the proposition below, which is due to V. Diekert and A. Muscholl. Essentially, it states that for a given partially commutative monoid $\mathbb{T}$, there are finitely many ways to take the product of two words (written in DM-normal form) in $\mathbb{T}$ to DM-normal form. More precisely, given a trace monoid $\mathbb{T}$ there exists a global bound k such that the words u and v can be written as a product of k subwords in such a way that the product $w = uv$ written in the normal form is obtained by concatenating these subwords in some order.

PROPOSITION 3.18 (Theorem 8, [**DM06**]). *Let u, v, w be words in DM-normal form in the trace monoid $\mathbb{T}$, such that $uv = w$ in $\mathbb{T}$. Then there exists a positive integer k bounded above by a computable function of $\mathfrak{r}$ (recall that $\mathfrak{r} = |\mathcal{A}|$) such that*

$$
\begin{aligned}
u &\doteq u_1 \cdots u_k \\
v &\doteq v_{j_1} \cdots v_{j_k} \\
w &\doteq u_1 v_1 \cdots u_k v_k
\end{aligned}
$$

for some permutation $(j_1, \ldots, j_k)$ of $\{1, \ldots, k\}$ and words u_i, v_j written in DM-normal form satisfying

$$u_i \leftrightarrows v_j \text{ for all } i > j \text{ and } v_{j_p} \leftrightarrows v_{j_q} \text{ for all } p, q \text{ such that } (j_p - j_q)(p - q) < 0$$

COROLLARY 3.19. *Let $w, w_1 \ldots w_l$ be words in DM-normal form in the trace monoid $\mathbb{T}$, such that $w_1 \ldots w_l = w$ in $\mathbb{T}$. Then there exists a positive integer $\mathfrak{k} = \mathfrak{k}(l, \mathfrak{r})$ bounded above by a computable function of l and $\mathfrak{r}$ such that*

$$
\begin{aligned}
w_i &\doteq w_{i,\varsigma_i(1)} \cdots w_{i,\varsigma_i(\mathfrak{k})}, \quad i = 1, \ldots, l; \\
w &\doteq w_{1,1} w_{2,1} \ldots w_{l,1} \ldots w_{1,\mathfrak{k}} \ldots w_{l,\mathfrak{k}}
\end{aligned}
$$

for some permutations ς_i, $i = 1, \ldots, l$ of $\{1, \ldots, \mathfrak{k}\}$ and some words $w_{i,j}$ written in DM-normal form satisfying

$w_{i_1, j_1} \leftrightarrows w_{i_2, j_2}$ *for all $i_1 > i_2$ and $j_1 < j_2$, and*

$w_{i, \varsigma_i(j_1)} \leftrightarrows w_{i, \varsigma_i(j_2)}$ *for all j_1, j_2 such that $(j_1 - j_2)(\varsigma_i(j_1) - \varsigma_i(j_2)) < 0$.*

PROOF. The proof is by induction on l and is left to the reader. □

3.3.1. Generalised equations over $\mathbb{F}$ and generalised equations over $\mathbb{T}$.
Let $\Omega' = \langle \Upsilon', \Re_{\Upsilon'} \rangle$ be a generalised equation over $\mathbb{T}$ and let $\Upsilon' = \{L_i = R_i \mid i = 1, \ldots, m\}$. To every word L_i (correspondingly, R_i) (in the alphabet h') we associate a set $C(L_i)$ (correspondingly, $C(R_i)$) constructed below.

We explain the idea behind this construction. By Corollary 3.19, for every solution H the word $L_i(H)$ (correspondingly, $R_i(H)$) can be taken to the DM-normal form by using appropriate permutations ς_i and setting some of the words $w_{i,\varsigma_i(j)}$ to be trivial. The system of equations in c, $c \in C(L_i)$ (correspondingly, $d \in C(R_i)$), see below for definition, is obtained from the equations

$$w_i \doteq w_{i,\varsigma_i(1)} \cdots w_{i,\varsigma_i(k)},$$

see Corollary 3.19, setting some of the words $w_{i,\varsigma_i(j)}$ to be trivial and applying the appropriate permutations ς_i.

The word $NF_c(L_i) \in c$, $c \in C(L_i)$ (correspondingly, $NF_d(R_i)$, $d \in C(R_i)$) is the DM-normal form of the word L_i (correspondingly, R_i) obtained from the equation

$$w \doteq w_{1,1} w_{2,1} \cdots w_{l,1} \cdots w_{1,\mathfrak{k}} \cdots w_{l,\mathfrak{k}},$$

see Corollary 3.19, setting some of the words $w_{i,\varsigma_i(j)}$ to be trivial and applying the appropriate permutations ς_i.

The relations $\mathcal{R}_c(L_i)$ (correspondingly, $\mathcal{R}_d(R_i)$) are the ones induced by Corollary 3.19.

We treat the constant items h'_{const} (see below) separately, since, for every solution every item in h'_{const} corresponds to an element from $\mathcal{A}^{\pm 1}$ and, therefore does not need to be subdivided to be taken to the normal form.

The formalisation of the above ideas is rather involved and technical. We refer the reader to Section 3.4 for an example.

Let

$$h'_{\text{const}} = \left\{ h'_i \in h' \;\middle|\; \begin{array}{l} \text{there is a coefficient equation of the form} \\ h'_i = a \text{ in } \Upsilon' \text{ for some } a \in \mathcal{A} \end{array} \right\}$$

Given L_i, an element c of $C(L_i)$ is a triple:

(system of equations in $h^{(L_i)} \cup h'$; a word $NF_c(L_i)$ in $h^{(L_i)}$;

a symmetric subset $\mathcal{R}_c(L_i)$ of $h^{(L_i)} \times h^{(L_i)}$).

Suppose first that the length of L_i (treated as a word in h') is greater than 1. Then the system of equations is defined as follows:

$$\left\{ \begin{array}{ll} h'^\epsilon_j = h^{(L_i)}_{j,\varsigma_{c,j}(i_{c,j,1})} \cdots h^{(L_i)}_{j,\varsigma_{c,j}(i_{c,j,k_{c,j}})}, & \text{for each } h'^\epsilon_j \text{ in } L_i,\, h'_j \notin h'_{\text{const}},\, \epsilon = \pm 1, \\ h'^\epsilon_j = h^{(L_i)}_{j,1}, & \text{for each } h'^\epsilon_j \text{ in } L_i,\, h'_j \in h'_{\text{const}},\, \epsilon = \pm 1 \end{array} \right\},$$

where $1 \leq i_{c,j,1} < i_{c,j,2} < \cdots < i_{c,j,k_{c,j}} \leq \mathfrak{k}$, $\mathfrak{k} = \mathfrak{k}(|L_i|, \mathfrak{r})$, $\varsigma_{c,j}$ is a permutation on $k_{c,j}$ symbols, $h^{(L_i)}_{j,\varsigma_{c,j}(i_{c,j,1})}, \ldots, h^{(L_i)}_{j,\varsigma_{c,j}(i_{c,j,k_{c,j}})} \in h^{(L_i)}$, see Corollary 3.19.

The word $NF_c(L_i)$ is a product of all the variables $h^{(L_i)}$. The variable $h^{(L_i)}_{j_1,k_1}$ is to the left of $h^{(L_i)}_{j_2,k_2}$ in $NF_c(L_i)$ if and only if (j_1, k_1) precedes (j_2, k_2) in the right lexicographical order.

The symmetric subset $\mathcal{R}_c(L_i)$ of $h^{(L_i)} \times h^{(L_i)}$ is defined as follows (cf. Corollary 3.19):

3.3. FROM PARTIALLY COMMUTATIVE MONOIDS TO FREE MONOIDS

- For all $j_1 > j_2$ and $k_1 < k_2$ we set

$$\left(h_{j_1,k_1}^{(L_i)}, h_{j_2,k_2}^{(L_i)}\right) \in \mathcal{R}_c(L_i);$$

- For all k_1, k_2 such that $(k_1 - k_2)(\varsigma_{c,j}(k_1) - \varsigma_{c,j}(k_2)) < 0$ we set

$$\left(h_{j,\varsigma_{c,j}(k_1)}^{(L_i)}, h_{j,\varsigma_{c,j}(k_2)}^{(L_i)}\right) \in \mathcal{R}_c(L_i).$$

Suppose now that $L_i = h'_j$, then the system of equations is: $\{h'_j = h_{j,1}^{(L_i)}\}$, the word $NF_c(L_i) = h_{j,1}^{(L_i)}$ and $\mathcal{R}_c(L_i) = \emptyset$.

If L_i is just a constant $a_{i_1}^{\pm 1}$, we define $C(L_i) = \{c_i\}$, $c_i = (\emptyset, a_{i_1}^{\pm 1}, \emptyset)$.
The construction of the set $C(R_i)$ is analogous.

REMARK 3.20. Notice that the sets $C(L_i)$ and of $C(R_i)$ can be effectively constructed and that their cardinality is bounded above by a computable function of $|L_i|$, $|R_i|$ and $\mathfrak{r}$.

Given a generalised equation $\Omega' = \langle \Upsilon', \Re_{\Upsilon'} \rangle$ over $\mathbb{T}$, where $\Upsilon' = \{L_1 = R_1, \ldots, L_m = R_m\}$ we construct a finite set of generalised equations $\mathcal{GE}(\Omega')$ over $\mathbb{F}$

$$\mathcal{GE}(\Omega') = \{\Omega_T \mid T = (c_1, \ldots, c_m, d_1, \ldots, d_m), \text{ where } c_i \in C(L_i), d_i \in C(R_i)\},$$

and $\Omega_T = \langle \Upsilon_T, \Re_{\Upsilon_T} \rangle$ is constructed as follows. The generalised equation Υ_T in variables $h = \bigcup_{i=1}^{m} \left(h^{(L_i)} \cup h^{(R_i)}\right)$ consists of the following equations:

(a) Equating the normal forms of L_i and R_i:

$$NF_{c_i}(L_i) = NF_{d_i}(R_i), \text{ for } i = 1, \ldots, m.$$

(b) Equating different decompositions (as words in the h's) of the same variable h'_j of Υ':

- for any two occurrences h'^{ϵ}_j and h'^{δ}_j of h'_j in equations of c_{i_1} and c_{i_2} respectively, $\epsilon, \delta \in \{-1, 1\}$, write:

$$\left(h_{j,\varsigma_{c_{i_1}},j(i_{c_{i_1},j,1})}^{(L_{i_1})} \cdots h_{j,\varsigma_{c_{i_1}},j(i_{c_{i_1},j,k_{c_{i_1}},j})}^{(L_{i_1})}\right)^{\epsilon} =$$
$$= \left(h_{j,\varsigma_{c_{i_2}},j(i_{c_{i_2},j,1})}^{(L_{i_2})} \cdots h_{j,\varsigma_{c_{i_2}},j(i_{c_{i_2},j,k_{c_{i_2}},j})}^{(L_{i_2})}\right)^{\delta};$$

- for any two occurrences h'^{ϵ}_j and h'^{δ}_j of h'_j in equations of c_{i_1} and d_{i_2} respectively, $\epsilon, \delta \in \{-1, 1\}$, write:

$$\left(h_{j,\varsigma_{c_{i_1}},j(i_{c_{i_1},j,1})}^{(L_{i_1})} \cdots h_{j,\varsigma_{c_{i_1}},j(i_{c_{i_1},j,k_{c_{i_1}},j})}^{(L_{i_1})}\right)^{\epsilon} =$$
$$= \left(h_{j,\varsigma_{d_{i_2}},j(i_{d_{i_2},j,1})}^{(R_{i_2})} \cdots h_{j,\varsigma_{d_{i_2}},j(i_{d_{i_2},j,k_{d_{i_2}},j})}^{(R_{i_2})}\right)^{\delta};$$

- for any two occurrences $h_j'^{\epsilon}$ and $h_j'^{\delta}$ of h_j' in equations of d_{i_1} and d_{i_2} respectively, $\epsilon, \delta \in \{-1, 1\}$, write:

$$\left(h_{j,\varsigma d_{i_1},j(i_{d_{i_1},j,1})}^{(R_{i_1})} \cdots h_{j,\varsigma d_{i_1},j(i_{d_{i_1},j,k_{d_{i_1}},j})}^{(R_{i_1})} \right)^{\epsilon} =$$

$$= \left(h_{j,\varsigma d_{i_2},j(i_{d_{i_2},j,1})}^{(R_{i_2})} \cdots h_{j,\varsigma d_{i_2},j(i_{d_{i_2},j,k_{d_{i_2}},j})}^{(R_{i_2})} \right)^{\delta},$$

The set of boundary connections of Υ_T is empty.

We now define $\Re_{\Upsilon_T}$. Set

- $\Re_{\Upsilon_T}\left(h_{j_1,k_1}^{(L_{i_1})}, h_{j_2,k_2}^{(L_{i_2})}\right)$, $\Re_{\Upsilon_T}\left(h_{j_1,k_1}^{(R_{i_1})}, h_{j_2,k_2}^{(L_{i_2})}\right)$, $\Re_{\Upsilon_T}\left(h_{j_1,k_1}^{(R_{i_1})}, h_{j_2,k_2}^{(R_{i_2})}\right)$ if $\Re_{\Upsilon'}(h_{j_1}', h_{j_2}')$;
- $\Re_{\Upsilon_T}\left(h_{j_1,k_1}^{(L_{i_1})}, h_{j_2,k_2}^{(L_{i_2})}\right)$ if $\left(h_{j_1,k_1}^{(L_{i_1})}, h_{j_2,k_2}^{(L_{i_2})}\right) \in \mathcal{R}_{c_i}(L_i)$ for some i;
- $\Re_{\Upsilon_T}\left(h_{j_1,k_1}^{(R_{i_1})}, h_{j_2,k_2}^{(R_{i_2})}\right)$ if $\left(h_{j_1,k_1}^{(R_{i_1})}, h_{j_2,k_2}^{(R_{i_2})}\right) \in \mathcal{R}_{d_i}(R_i)$ for some i.

We define $\Re_{\Upsilon_T}$ to be the minimal subset of $h \times h$ that is symmetric and satisfies condition $(\star)$ of Definition 3.4. This defines a combinatorial generalised equation, see Lemma 3.3.

3.3.2. Coordinate groups of generalised equations over $\mathbb{T}$ and over $\mathbb{F}$. We now explain the relation between the coordinate group of the generalised equation Ω' over $\mathbb{T}$ and the coordinate group of the generalised equation Ω_T over $\mathbb{F}$. For any variable $h_j' \in h'$ we choose an arbitrary equation in c_i, $c_i \in T$ (or d_i, $d_i \in T$) of the form $h_j'^{\epsilon} = h_{j,\varsigma c_i,j(i_{c,j,1})}^{(L_i)} \cdots h_{j,\varsigma c_i,j(i_{c_i,j,k_{c_i},j})}^{(L_i)}$ and define a word $P_{h_j'}(h, \mathcal{A}) \in \mathbb{G}[h]$ as follows:

$$P_{h_j'}(h, \mathcal{A}) = \left(h_{j,\varsigma c_i,j(i_{c_i,j,1})}^{(L_i)} \cdots h_{j,\varsigma c_i,j(i_{c_i,j,k_{c_i},j})}^{(L_i)} \right)^{\epsilon}.$$

The word $P_{h_j'}(h, \mathcal{A})$ depends on the choice of the equation in c_i and on i. It follows from the construction above, that the map

$$h' \to \mathbb{G}[h] \text{ defined by } h_j' \mapsto P_{h_j'}(h, \mathcal{A})$$

gives rise to a $\mathbb{G}$-homomorphism $\pi_{\Omega_T} : \mathbb{G}_{R(\Omega'^*)} \to \mathbb{G}_{R(\Omega_T^*)}$. Observe that the image $\pi_{\Omega_T}(h_j') \in \mathbb{G}_{R(\Omega_T^*)}$ does not depend on a particular choice of i and of the equation in c_i (or in d_i), since equations in Ω make all of them identical. Hence, π_{Ω_T} depends only on Ω_T.

If H is an arbitrary solution of the generalised equation Ω_T, then $H' = (P_{h_1'}(H, \mathcal{A}), \ldots, P_{h_{\rho'}'}(H, \mathcal{A}))$ is a solution of Ω'.

The converse also holds. From Corollary 3.19, the fact that DM-normal form is invariant with respect to inversion and definition of solution of a generalised equation, it follows that any solution H' of Ω' induces a solution H of Ω_T. Furthermore, if $H' = (H_1', \ldots, H_{\rho'}')$ then $H_j' = P_{h_j'}(H, \mathcal{A})$.

The following lemma shows that to describe the set of solutions of a constrained generalised equation over $\mathbb{T}$ is equivalent to describe the set of solutions of constrained generalised equations over $\mathbb{F}$. This lemma can be viewed as the second "divide" step of the process.

LEMMA 3.21. *Let Ω' be a generalised equation in variables $h' = \{h'_1, \ldots, h'_{\rho'}\}$ over $\mathbb{T}$, one can effectively construct a finite set $\mathcal{GE}(\Omega')$ of constrained generalised equations over $\mathbb{F}$ such that*
  (1) *if the set $\mathcal{GE}(\Omega')$ is empty, then Ω' has no solutions;*
  (2) *for each $\Omega(h) \in \mathcal{GE}(\Omega')$ and for each $h'_i \in h'$ one can effectively find a word $P_{h'_i}(h, \mathcal{A}) \in \mathbb{G}[h]$ of length at most $|h|$ such that the map $h'_i \mapsto P_{h'_i}(h, \mathcal{A})$ ($h'_i \in h'$) gives rise to a $\mathbb{G}$-homomorphism $\pi_\Omega : \mathbb{G}_{R(\Omega'^*)} \to \mathbb{G}_{R(\Omega^*)}$ (in particular, for every solution H of the generalised equation Ω one has that $P(H)$ is a solution of the generalised equation Ω', where $P(h) = (P_{h'_1}, \ldots, P_{h'_{\rho'}})$);*
  (3) *for any solution $H' \in \mathbb{G}^n$ of Ω' there exists $\Omega(h) \in \mathcal{GE}(\Omega')$ and a solution H of $\Omega(h)$ such that $H' = P(H)$, where $P(h) = (P_{h'_1}, \ldots, P_{h'_{\rho'}})$, and this equality holds in the free monoid $\mathbb{F}$.*

Combining Lemma 3.16 and Lemma 3.21 we get the following

COROLLARY 3.22. *In the notation of Lemmas 3.16 and 3.21 for any solution $W(\mathcal{A}) \in \mathbb{G}^n = \mathbb{G}(\mathcal{A})^n$ of the system $S(X, \mathcal{A}) = 1$ there exist a generalised equation $\Omega'(h') \in \mathcal{GE}'(S)$, a solution $H'(\mathcal{A})$ of $\Omega'(h')$, a generalised equation $\Omega(h) \in \mathcal{GE}(S)$, and a solution $H(\mathcal{A})$ of $\Omega(h)$ such that the following diagram commutes*

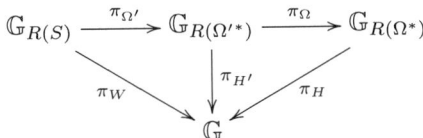

3.4. Example

The aim of Chapter 3 is to show that to a given finite system of equations $S = S(X, \mathcal{A}) = 1$ over a partially commutative group $\mathbb{G}$ one can associate a finite collection of (constrained) generalised equations $\mathcal{GE}(S)$ over $\mathbb{F}$ with coefficients from $\mathcal{A}^{\pm 1}$ such that for any solution $W(\mathcal{A}) \in \mathbb{G}^n$ of S there exists a generalised equation $\Omega(h) = \Omega$ over $\mathbb{F}$ and a solution $H(\mathcal{A})$ of Ω such that

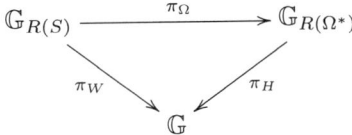

The family of solutions of the generalised equations from $\mathcal{GE}'(S)$ describes all solutions of the system $S(X, \mathcal{A}) = 1$, see Lemma 3.16.

In this section for a particular equation over a given partially commutative group $\mathbb{G}$ and one of its solutions, following the exposition of Chapter 3, we first construct the constrained generalised equation over $\mathbb{T}$ and then the generalised equation over $\mathbb{F}$ such that the above diagram commutes.

Let $\mathbb{G}$ be the free partially commutative group whose underlying commutation graph is a pentagon, see Figure 1. Let $S(x, y, z, \mathcal{A}) = \{xyzy^{-1}x^{-1}z^{-1}ebe^{-1}b^{-1} = 1\}$ be a system consisting of a single equation over $\mathbb{G}$ in variables x, y and z. Since the word

$$S(bac, c^{-1}a^{-1}d, e) = bac\,c^{-1}a^{-1}d\,e\,d^{-1}ac\,c^{-1}a^{-1}b^{-1}\,e^{-1}\,ebe^{-1}b^{-1}$$

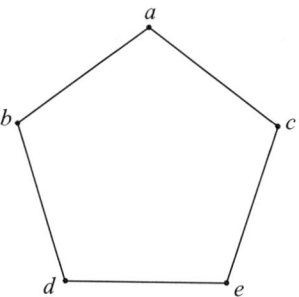

FIGURE 1. The commutation graph of $\mathbb{G}$.

is trivial in $\mathbb{G}$, the tuple $x = bac$, $y = c^{-1}a^{-1}d$, $z = e$ is a solution of S. Construct a van Kampen diagram $\mathcal{D}$ for the word $S(bac, c^{-1}a^{-1}d, e) = 1$ and consider the underlying cancellation scheme as in Proposition 2.7, see Figure 2.

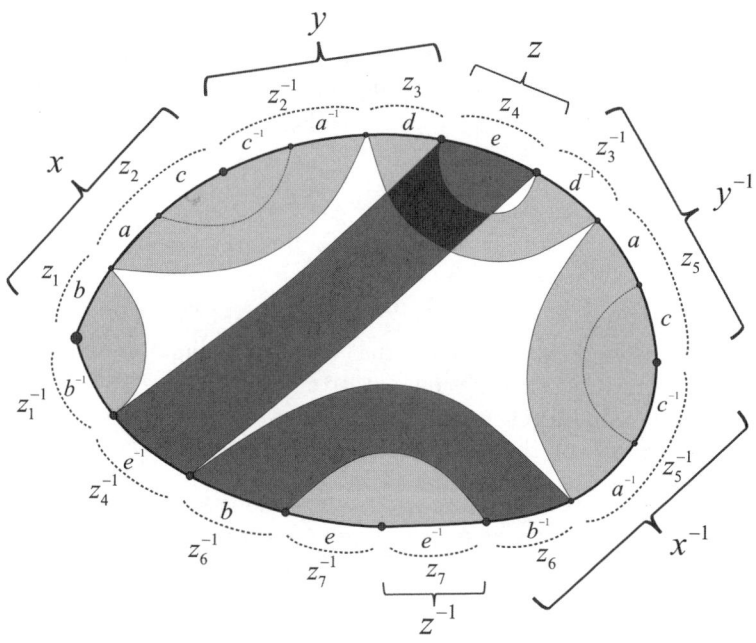

FIGURE 2. Cancellation scheme for the solution $x = bac$, $y = c^{-1}a^{-1}d$, $z = e$ of S.

Content of Section 3.2.1. Write the equation $S(x, y, z, \mathcal{A})$ in the form $xyzy^{-1}x^{-1}z^{-1}ebe^{-1}b^{-1} = r_{1,1} \ldots r_{1,10} = 1$, see Equation (3.1).

By the cancellation scheme shown on Figure 2, we construct a partition table $\mathcal{T}$. To every end of a band on Figure 2, we associate a variable z_i or its inverse z_i^{-1}. We can write the variables x, y, z as words in the variables z_i's. Two variables z_i and z_j commute if and only if the corresponding bands cross. Thus, the partition table associated to the equation $S(x, y, z, \mathcal{A})$ and its solution $x = bac$, $y = c^{-1}a^{-1}d$,

3.4. EXAMPLE

$z = e$ is given below

$$\mathcal{T} = (\{V_{1i}(z_1, \ldots, z_7)\}), \quad \mathbb{H} = \langle \mathbb{G}, z_1, \ldots, z_7 \mid [z_3, z_4] = 1 \rangle, \quad i = 1, \ldots, 10$$

$V_{1,1} = z_1 z_2;$ $V_{1,4} = z_3^{-1} z_5;$ $V_{1,7} = z_7^{-1}$ $V_{1,9} = z_4^{-1};$
$V_{1,2} = z_2^{-1} z_3;$ $V_{1,5} = z_5^{-1} z_6;$ $V_{1,8} = z_6^{-1}$ $V_{1,10} = z_1^{-1}.$
$V_{1,3} = z_4';$ $V_{1,6} = z_7;$

In the above notation, it is easy to see that every variable $z_i^{\pm 1}$, $i = 1, \ldots, 7$ occurs in the words $V_{1,j}$ precisely once, and that $V_1 = V_{1,1} \cdots V_{1,10} =_{\mathbb{H}} 1$.

Content of Section 3.2.2. We now construct the generalised equation $\Omega'_{\mathcal{T}}$ associated to the partition table $\mathcal{T}$. Consider the word $\mathcal{V}$ in the free monoid $\mathbb{F}(z_1, \ldots, z_7, z_1^{-1}, \ldots, z_7^{-1})$:

$$\mathcal{V} = V_1 = z_1 z_2 z_2^{-1} z_3 z_4 z_3^{-1} z_5 z_5^{-1} z_6 z_7 z_7^{-1} z_6^{-1} z_4^{-1} z_1^{-1} = y_1 \cdots y_{14}.$$

The variables of the generalised equation $\Omega'_{\mathcal{T}}$ are $h'_1, \ldots, h'_{14}$ (equivalently, the set of boundaries of the combinatorial generalised equation is $\{1, \ldots, 15\}$).

For every pair of distinct occurrences of z_i in $\mathcal{V}$ we construct a pair of dual variable bases and the corresponding basic equation. For example, $z_1 = y_1$ and $z_1^{-1} = y_{14}$. We introduce a pair of dual bases $\mu_{z_1,1}, \mu_{z_1,14}$ and $\Delta(\mu_{z_1,1}) = \mu_{z_1,14}$ so that $\alpha(\mu_{z_1,1}) = 1$, $\beta(\mu_{z_1,1}) = 2$, $\alpha(\Delta(\mu_{z_1,1})) = 14$, $\beta(\Delta(\mu_{z_1,1})) = 15$, $\varepsilon(\mu_{z_1,1}) = 1$, $\varepsilon(\mu_{z_1,14}) = -1$.

The corresponding basic equation is $h'_1 = h'^{-1}_{14}$, see Figure 3. The following table describes all basic equations thus constructed.

pair of occurrences	pair of dual bases	basic equation
$y_1 = z_1, y_{14} = z_1^{-1}$	$\Delta(\mu_{z_1,1}) = \mu_{z_1,14}$	$\{h'_1 = h'^{-1}_{14}\} = \{L_1 = R_1\}$
$y_2 = z_2, y_3 = z_2^{-1}$	$\Delta(\mu_{z_2,2}) = \mu_{z_2,3}$	$\{h'_2 = h'^{-1}_3\} = \{L_2 = R_2\}$
$y_4 = z_3, y_6 = z_3^{-1}$	$\Delta(\mu_{z_3,4}) = \mu_{z_3,6}$	$\{h'_4 = h'^{-1}_6\} = \{L_3 = R_3\}$
$y_5 = z_4, y_{13} = z_4^{-1}$	$\Delta(\mu_{z_4,5}) = \mu_{z_4,13}$	$\{h'_5 = h'^{-1}_{13}\} = \{L_4 = R_4\}$
$y_7 = z_5, y_8 = z_5^{-1}$	$\Delta(\mu_{z_5,7}) = \mu_{z_5,8}$	$\{h'_7 = h'^{-1}_8\} = \{L_5 = R_5\}$
$y_9 = z_6, y_{12} = z_6^{-1}$	$\Delta(\mu_{z_6,9}) = \mu_{z_6,12}$	$\{h'_9 = h'^{-1}_{12}\} = \{L_6 = R_6\}$
$y_{10} = z_7, y_{11} = z_7^{-1}$	$\Delta(\mu_{z_7,10}) = \mu_{z_7,11}$	$\{h'_{10} = h'^{-1}_{11}\} = \{L_7 = R_7\}$

For every pair of distinct occurrences of x (correspondingly, of y, or of z) in $r_{1,1} \cdots r_{1,10} = 1$, we construct a pair of dual variable bases and the corresponding basic equation. For example, $r_{1,1} = x$, $r_{1,5} = x^{-1}$. We introduce a pair of dual bases $\mu_{x,q_1}, \Delta(\mu_{x,q_1})$, where $q_1 = (1,1,1,5)$ so that $\alpha(\mu_{x,q_1}) = 1$, $\beta(\mu_{x,q_1}) = 3$, $\alpha(\Delta(\mu_{x,q_1})) = 8$, $\beta(\Delta(\mu_{x,q_1})) = 10$, $\varepsilon(\mu_{x,q_1}) = 1$, $\varepsilon(\Delta(\mu_{x,q_1})) = -1$.

The corresponding basic equation is $h'_1 h'_2 = (h'_8 h'_9)^{-1}$, see Figure 3. The following table describes all basic equations thus constructed.

pair of occurrences	pair of dual bases	basic equation
$r_{1,1} = x, r_{1,5} = x^{-1}$	$\Delta(\mu_{x,q_1}), \mu_{x,q_1}$	$\{h'_1 h'_2 = (h'_8 h'_9)^{-1}\} = \{L_8 = R_8\}$
$r_{1,2} = y, r_{1,4} = y^{-1}$	$\Delta(\mu_{y,q_2}), \mu_{y,q_2}$	$\{h'_3 h'_4 = (h'_6 h'_7)^{-1}\} = \{L_9 = R_9\}$
$r_{1,3} = z, r_{1,6} = z^{-1}$	$\Delta(\mu_{z,q_3}), \mu_{z,q_3}$	$\{h'_5 = h'^{-1}_{10}\} = \{L_{10} = R_{10}\}$

Here $q_1 = (1, 1, 1, 5)$, $q_2 = (1, 2, 1, 4)$, $q_3 = (1, 3, 1, 6)$.

For $r_{1j} = a \in \mathcal{A}^{\pm 1}$, we construct a constant base and a coefficient equation. For example, $r_{1,7} = e$. We introduce a constant base $\mu_{1,11}$ so that

$$\alpha(\mu_{1,11}) = 11, \quad \beta(\mu_{1,11}) = 12.$$

The corresponding coefficient equation is $h'_{11} = e$, see Figure 3. The following table describes all coefficient equations.

an occurrence	constant base	corresponding coefficient equation
$r_{1,11} = e$	$\mu_{1,11} = e$	$\{h'_{11} = e\} = \{L_{11} = R_{11}\}$
$r_{1,12} = b$	$\mu_{1,12} = b$	$\{h'_{12} = b\} = \{L_{12} = R_{12}\}$
$r_{1,13} = e^{-1}$	$\mu_{1,13} = e^{-1}$	$\{{h'_{13}}^{-1} = e\} = \{L_{13} = R_{13}\}$
$r_{1,14} = b^{-1}$	$\mu_{1,14} = b^{-1}$	$\{{h'_{14}}^{-1} = b\} = \{L_{14} = R_{14}\}$

The set of boundary connections of $\Upsilon'_\mathcal{T}$ is empty. This defines the generalised equation $\Upsilon'_\mathcal{T}$ shown on Figure 3.

The set of pairs (h'_i, h'_j) such that $[y_i, y_j] = 1$ in $\mathbb{H}$ and $y_i \neq y_j$ consists of:

$$\{(h'_4, h'_5), (h'_5, h'_4), (h'_5, h'_6), (h'_6, h'_5), (h'_4, h'_{13}), (h'_{13}, h'_4), (h'_6, h'_{13}), (h'_{13}, h'_6)\}.$$

To define the relation $\Re_{\Upsilon'_\mathcal{T}}$ we have to make sure that the set $\Re_{\Upsilon'_\mathcal{T}} \subseteq h' \times h'$ is symmetric and satisfies condition $(\star)$ of Definition 3.4. Since we have the equations $h'_5 = {h'_{10}}^{-1}$, $h'_{10} = {h'_{11}}^{-1}$, we get that

$$\Re_{\Upsilon'_\mathcal{T}} = \left\{ \begin{array}{l} (h'_4, h'_5), (h'_5, h'_6), (h'_4, h'_{10}), (h'_6, h'_{10}), \\ (h'_4, h'_{11}), (h'_6, h'_{11}), (h'_4, h'_{13}), (h'_6, h'_{13}) \end{array} \right\}.$$

Clearly, the relation $\Re_{\Upsilon'_\mathcal{T}}$ has to be symmetric, but, in this section, we further write only one of the two symmetric pairs. This defines the constrained generalised equation $\Omega'_\mathcal{T} = \langle \Upsilon'_\mathcal{T}, \Re_{\Upsilon'_\mathcal{T}} \rangle$.

We represent generalised equations graphically in the way shown on Figure 3.

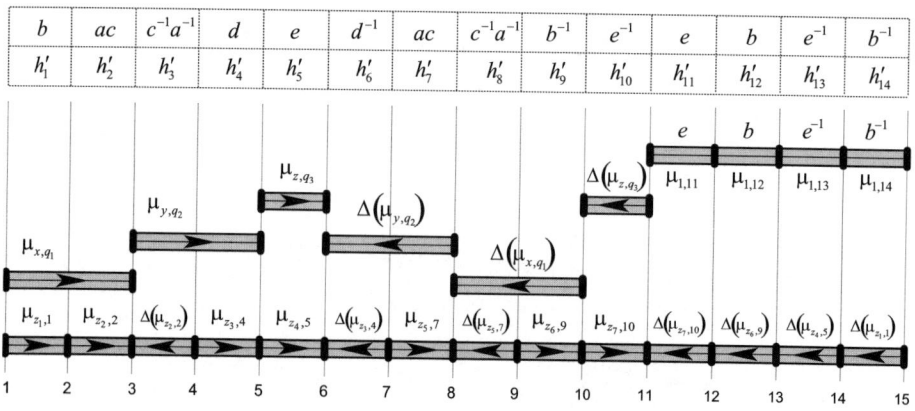

FIGURE 3. The generalised equation $\Omega'_\mathcal{T}$.

3.4. EXAMPLE

Content of Section 3.3.1. By Corollary 3.22 for every pair (Ω'_T, H') there exists a pair (Ω_T, H) such that the diagram given in Corollary 3.22 is commutative. We now construct the set $T = (c_1, \ldots, c_{14}, d_1, \ldots, d_{10})$ that defines a generalised equation Ω_T over $\mathbb{F}$. We construct the elements c_i of the sets $C(L_i)$ that correspond to the solution $x = bac$, $y = c^{-1}a^{-1}d$, $z = e$.

Recall that the partially commutative group $\mathbb{G}$ is defined by the commutation graph shown on Figure 1. The definition of DM-normal form in the partially commutative monoid $\mathbb{T}$ is given by induction on the number of thin clans. In our example thin clans of $\mathbb{T}$ are $\{a, a^{-1}\}$, $\{b, b^{-1}\}$, $\{c, c^{-1}\}$, $\{d, d^{-1}\}$ and $\{e, e^{-1}\}$.

Consider the DM-normal form in $\mathbb{T}$ with respect to the thin clan $\{a, a^{-1}\}$. Note that for $L_8 = h'_1 h'_2$, we have that $L_8(bac, c^{-1}a^{-1}d, e) = bac$ is not written in the DM-normal form. We introduce a new variable and write $h'_1 = h^{(L_8)}_{1,2}$, $h'_2 = h^{(L_8)}_{2,1} h^{(L_8)}_{2,2}$ (here the corresponding permutation $\varsigma_{c,2}$ on $\{1,2\}$ is trivial) and $NF_{c_8}(L_8) = h^{(L_8)}_{2,1} h^{(L_8)}_{1,2} h^{(L_8)}_{2,2}$. Notice that $NF_{c_8}(L_8)(bac, c^{-1}a^{-1}d, e) = abc$ is in the DM-normal form.

Below we describe all the elements c_i for the solution $x = bac$, $y = c^{-1}a^{-1}d$, $z = e$.

$$C(L_1) \ni c_1 = \left(\left\{h'_1 = h^{(L_1)}_{1,1}\right\}; h^{(L_1)}_{1,1}; \emptyset\right);$$

$$C(L_2) \ni c_2 = \left(\left\{h'_2 = h^{(L_2)}_{2,1}\right\}; h^{(L_2)}_{2,1}; \emptyset\right);$$

$$C(L_3) \ni c_3 = \left(\left\{h'_4 = h^{(L_3)}_{4,1}\right\}; h^{(L_3)}_{4,1}; \emptyset\right);$$

$$C(L_4) \ni c_4 = \left(\left\{h'_5 = h^{(L_4)}_{5,1}\right\}; h^{(L_4)}_{5,1}; \emptyset\right);$$

$$C(L_5) \ni c_5 = \left(\left\{h'_7 = h^{(L_5)}_{7,1}\right\}; h^{(L_5)}_{7,1}; \emptyset\right);$$

$$C(L_6) \ni c_6 = \left(\left\{h'_9 = h^{(L_6)}_{9,1}\right\}; h^{(L_6)}_{9,1}; \emptyset\right);$$

$$C(L_7) \ni c_7 = \left(\left\{h'_{10} = h^{(L_7)}_{10,1}\right\}; h^{(L_7)}_{10,1}; \emptyset\right);$$

$$C(L_8) \ni c_8 = \left(\begin{array}{l}\left\{h'_1 = h^{(L_8)}_{1,2}, h'_2 = h^{(L_8)}_{2,1} h^{(L_8)}_{2,2}\right\}; \\ h^{(L_8)}_{2,1} h^{(L_8)}_{1,2} h^{(L_8)}_{2,2}; \mathcal{R}_c(L_8) = \left\{\left(h^{(L_8)}_{1,2}, h^{(L_8)}_{2,1}\right)\right\}\end{array}\right);$$

$$C(L_9) \ni c_9 = \left(\left\{h'_3 = h^{(L_9)}_{3,1}, h'_4 = h^{(L_9)}_{4,1}\right\}; h^{(L_9)}_{3,1} h^{(L_9)}_{4,1}; \emptyset\right)$$

$$C(L_{10}) \ni c_{10} = \left(\left\{h'_5 = h^{(L_{10})}_{5,1}\right\}; h^{(L_{10})}_{5,1}; \emptyset\right);$$

$$C(L_{11}) \ni c_{11} = \left(\left\{h'_{11} = h^{(L_{11})}_{11,1}\right\}; h^{(L_{11})}_{11,1}; \emptyset\right);$$

$$C(L_{12}) \ni c_{12} = \left(\left\{h'_{12} = h^{(L_{12})}_{12,1}\right\}; h^{(L_{12})}_{12,1}; \emptyset\right);$$

$$C(L_{13}) \ni c_{13} = \left(\left\{{h'_{13}}^{-1} = h^{(L_{13})}_{13,1}\right\}; h^{(L_{13})}_{13,1}; \emptyset\right)$$

$$C(L_{14}) \ni c_{14} = \left(\left\{{h'_{14}}^{-1} = h^{(L_1)}_{14,1}\right\}; h^{(L_{14})}_{14,1}; \emptyset\right).$$

46   3. FROM EQUATIONS OVER G TO GENERALISED EQUATIONS OVER F

Analogously, we describe all the elements d_i for the solution $x = bac$, $y = c^{-1}a^{-1}d$, $z = e$.

$$C(R_1) \ni d_1 = \left(\left\{h'_{14} = h^{(R_1)}_{14,1}\right\}; h^{(R_1)}_{14,1}; \emptyset\right);$$
$$C(R_2) \ni d_2 = \left(\left\{h'_3 = h^{(R_2)}_{3,1}\right\}; h^{(R_2)}_{3,1}; \emptyset\right);$$
$$C(R_3) \ni d_3 = \left(\left\{h'_6 = h^{(R_3)}_{6,1}\right\}; h^{(R_3)}_{6,1}; \emptyset\right);$$
$$C(R_4) \ni d_4 = \left(\left\{h'_{13} = h^{(R_4)}_{13,1}\right\}; h^{(R_4)}_{13,1}; \emptyset\right);$$
$$C(R_5) \ni d_5 = \left(\left\{h'_8 = h^{(R_5)}_{8,1}\right\}; h^{(R_5)}_{8,1}; \emptyset\right);$$
$$C(R_6) \ni d_6 = \left(\left\{h'_{12} = h^{(R_6)}_{12,1}\right\}; h^{(R_6)}_{12,1}; \emptyset\right);$$
$$C(R_7) \ni d_7 = \left(\left\{h'_{11} = h^{(R_7)}_{11,1}\right\}; h^{(R_7)}_{11,1}; \emptyset\right);$$
$$C(R_8) \ni d_8 = \left(\begin{array}{c}\left\{{h'_9}^{-1} = h^{(R_8)}_{9,1}, {h'_8}^{-1} = h^{(R_8)}_{8,1}h^{(R_8)}_{8,2}\right\}; \\ h^{(R_8)}_{8,1}h^{(R_8)}_{9,1}h^{(R_8)}_{8,2}; \mathcal{R}_c(R_8) = \left\{\left(h^{(R_8)}_{8,1}, h^{(R_8)}_{9,1}\right)\right\}\end{array}\right);$$
$$C(R_9) \ni d_9 = \left(\left\{{h'_7}^{-1} = h^{(R_9)}_{7,1}, {h'_6}^{-1} = h^{(R_9)}_{6,1}\right\}; h^{(R_9)}_{6,1}h^{(R_9)}_{7,1}; \emptyset\right);$$
$$C(R_{10}) \ni d_{10} = \left(\left\{{h'_{10}}^{-1} = h^{(R_{10})}_{10,1}\right\}; h^{(R_{10})}_{10,1}; \emptyset\right);$$
$$C(R_{11}) \ni d_{11} = (\emptyset; e; \emptyset); \quad C(R_{12}) \ni d_{12} = (\emptyset; b; \emptyset);$$
$$C(R_{13}) \ni d_{13} = (\emptyset; e^{-1}; \emptyset); \quad C(R_{14}) \ni d_{14} = (\emptyset; b^{-1}; \emptyset).$$

For the tuple $T = (c_1, \ldots, c_{14}, d_1, \ldots, d_{14})$ we construct the generalised equation Ω_T. Equations corresponding to $NF_{c_i}(L_i) = NF_{d_i}(R_i)$, $i = 1, \ldots, 14$ are:

$h^{(L_1)}_{1,1} = h^{(R_1)}_{14,1}$; $\quad h^{(L_6)}_{9,1} = h^{(R_6)}_{12,1}$; $\qquad\qquad h^{(L_{11})}_{11,1} = e$;

$h^{(L_2)}_{2,1} = h^{(R_2)}_{3,1}$; $\quad h^{(L_7)}_{10,1} = h^{(R_7)}_{11,1}$; $\qquad\qquad h^{(L_{12})}_{12,1} = b$;

$h^{(L_3)}_{4,1} = h^{(R_3)}_{6,1}$; $\quad h^{(L_8)}_{2,1}h^{(L_8)}_{1,2}h^{(L_8)}_{2,2} = h^{(R_8)}_{8,1}h^{(R_8)}_{9,1}h^{(R_8)}_{8,2}$; $\quad h^{(L_{13})}_{13,1} = e^{-1}$;

$h^{(L_4)}_{5,1} = h^{(R_4)}_{13,1}$; $\quad h^{(L_9)}_{3,1}h^{(L_9)}_{4,1} = h^{(R_9)}_{6,1}h^{(R_9)}_{7,1}$; $\qquad h^{(L_{14})}_{14,1} = b^{-1}$.

$h^{(L_5)}_{7,1} = h^{(R_5)}_{8,1}$; $\quad h^{(L_{10})}_{5,1} = h^{(R_{10})}_{10,10,1}$;

Equations that equate different decompositions of the same variable h'_j are obtained as follows:

h'_1: from L_1, L_8, we get $h^{(L_1)}_{1,1} = h^{(L_8)}_{1,2}$;

h'_2: from L_2, L_8, we get $h^{(L_2)}_{2,1} = h^{(L_8)}_{2,1}h^{(L_8)}_{2,2}$;

h'_3: from L_9, R_2, we get $h^{(L_9)}_{3,1} = {h^{(R_2)}_{3,1}}^{-1}$;

h'_4: from L_3, L_9, we get $h^{(L_3)}_{4,1} = h^{(L_9)}_{4,1}$;

h'_5: from L_4, L_{10}, we get $h^{(L_4)}_{5,1} = h^{(L_{10})}_{5,1}$;

h'_6: from R_3, R_9, we get $h^{(R_3)}_{6,1} = {h^{(R_9)}_{6,1}}^{-1}$;

h'_7: from L_5, R_9, we get $h^{(L_5)}_{7,1} = {h^{(R_9)}_{7,1}}^{-1}$;

h'_8: from R_5, R_8, we get $h^{(R_5)}_{8,1} = {h^{(R_8)}_{8,2}}^{-1}{h^{(R_8)}_{8,1}}^{-1}$;

h'_9: from L_6, R_8, we get $h^{(L_6)}_{9,1} = {h^{(R_8)}_{9,1}}^{-1}$;

h'_{10}: from L_7, R_{10}, we get $h^{(L_7)}_{10,1} = {h^{(R_{10})}_{10,1}}^{-1}$;

h'_{11}: from L_{11}, R_7, we get $h^{(L_{11})}_{11,1} = h^{(R_7)}_{11,1}$;

h'_{12}: from L_{12}, R_6, we get $h^{(L_{12})}_{12,1} = h^{(R_6)}_{12,1}$;

h'_{13}: from L_{13}, R_4, we get $h_{13,1}^{(L_{13})}{}^{-1} = h_{13,1}^{(R_4)}$;

h'_{14}: from L_{14}, R_1, we get $h_{14,1}^{(L_{14})} - 1 = h_{14,1}^{(R_1)}$.

This defines the generalised equation Υ_T (by the definition, the set of boundary connections of Υ_T is empty).

Set

- $\Re_{\Upsilon_T}\left(h_{j_1,k_1}^{(L_{i_1})}, h_{j_2,k_2}^{(L_{i_2})}\right)$, $\Re_{\Upsilon_T}\left(h_{j_1,k_1}^{(R_{i_1})}, h_{j_2,k_2}^{(L_{i_2})}\right)$, $\Re_{\Upsilon_T}\left(h_{j_1,k_1}^{(R_{i_1})}, h_{j_2,k_2}^{(R_{i_2})}\right)$ if $\Re_{\Upsilon'}(h'_{j_1}, h'_{j_2})$;

- $\Re_{\Upsilon_T}\left(h_{j_1,k_1}^{(L_{i_1})}, h_{j_2,k_2}^{(L_{i_2})}\right)$ if $\left(h_{j_1,k_1}^{(L_{i_1})}, h_{j_2,k_2}^{(L_{i_1})}\right) \in \mathcal{R}_{c_i}(L_i)$ for some i;

- $\Re_{\Upsilon_T}\left(h_{j_1,k_1}^{(R_{i_1})}, h_{j_2,k_2}^{(L_{i_2})}\right)$ if $\left(h_{j_1,k_1}^{(R_{i_1})}, h_{j_2,k_2}^{(L_{i_1})}\right) \in \mathcal{R}_{d_i}(R_i)$ for some i.

We define $\Re_{\Upsilon_T}$ to be the minimal subset of $h \times h$ that is symmetric and satisfies condition $(\star)$ of Definition 3.4.

REMARK 3.23. Observe, that the initial system of equations $S(x, y, z, \mathcal{A})$ contains a single equation in three variables. The generalised equation Ω'_T is a system of equations in 14 variables. The generalised equation Ω_T is a system of equations in at least 30 variables. (Note that for Ω_T, we constructed the system of equations over the monoid $\mathbb{F}$. We still should have constructed a combinatorial generalised equation associated to this system, see Lemma 3.3.)

Notice that, the above considered example describes only one generalised equation (which was traced by a particular solution of $S(x, y, z, \mathcal{A})$). Consideration of all the *finite* collection of the generalised equations corresponding to the "simple" system of equations $S(x, y, z, \mathcal{A})$, can only be done on a computer.

This picture is, in fact, general. The size of the systems grows dramatically and makes difficult to work with examples.

CHAPTER 4

The process: construction of the tree T

> *"Es werden aber bei uns in der Regel keine aussichtslosen Prozesse geführt."*
> Franz Kafka, "Der Prozess"

In the previous chapter we reduced the study of the set of solutions of a system of equations over a partially commutative group to the study of solutions of constrained generalised equations over a free monoid. In order to describe the solutions of constrained generalised equations over a free monoid, in this chapter we describe a branching rewriting process for constrained generalised equations.

In his important papers [**Mak77**] and [**Mak82**], G. Makanin devised a process for proving that the compatibility problem of systems of equations over a free monoid (over a free group) is decidable. This process was later developed by A. Razborov. In his work [**Raz85**], [**Raz87**], gave a complete description of all solutions of a system of equations. A further step was made by O. Kharlampovich and A. Miasnikov. In [**KhM05a**], in particular, the authors extend Razborov's result to systems of equations with parameters. In [**KhM05b**] the authors establish a correspondence between the process and the JSJ decomposition of fully residually free groups.

In another direction, Makanin's result (on decidability of equations over a free monoid) was developed by K.Schulz, see [**Sch90**], who proved that the compatibility problem of equations with regular constraints over a free monoid is decidable.

It turns out to be that the compatibility problem of equations over numerous groups and monoids can be reduced to the compatibility problem of equations over a free monoid with constraints. This technique turned out to be rather fruitful. G. Makanin was the first to do such a reduction. In [**Mak82**], in particular, he reduced the compatibility problem of equations over a free group to the decidability of the compatibility problem for free monoids. Later V. Diekert, C. Gutiérrez and C. Hagenah, see [**DGH01**], reduced the compatibility problem of systems of equations over a free group with rational constraints to compatibility problem of equations with regular constraints over a free monoid.

The reduction of compatibility problem for hyperbolic groups to free group was made in [**RS95**] by E. Rips and Z. Sela; for relatively hyperbolic groups with virtually abelian parabolic subgroups in [**Dahm09**] by F. Dahmani, for HNN-extensions with finite associated subgroups and for amalgamated products with finite amalgamated subgroups in [**LS08**] by M. Lohrey and G. Sénizergues, for partially commutative monoids in [**Mat97**] by Yu. Matiasevich, for partially commutative groups in [**DM06**] by V. Diekert and A. Muscholl, for graph product of groups in [**DL04**] by V. Diekert and M. Lohrey.

The complexity of Makanin's algorithm has received a great deal of attention. The best result about arbitrary systems of equations over monoids is due to W. Plandowski. In a series of two papers [**Pl99a, Pl99b**] he gave a new approach to the compatibility problem of systems of equations over a free monoid and showed that this problem is in PSPACE. An important ingredient of Plandowski's method is data compression in terms of exponential expressions. This approach was further extended by Diekert, Gutiérrez and Hagenah, see [**DGH01**] to systems of equations over free groups. Recently, O. Kharlampovich, I. Lysënok, A. Myasnikov and N. Touikan have shown that solving quadratic equations over free groups is NP-complete, [**KhLMT08**].

Another important development of the ideas of Makanin is due to E. Rips and is now known as the Rips' machine. In his work Rips interprets Makanin's algorithm in terms of partial isometries of real intervals, which leads him to a classification theorem of finitely generated groups that act freely on $\mathbb{R}$-trees. A complete proof of Rips' theorem was given by D. Gaboriau, G. Levitt, and F. Paulin, see [**GLP94**], and, independently, by M. Bestvina and M. Feighn, see [**BF95**], who also generalised Rips' result to give a classification theorem of groups that have a stable action on $\mathbb{R}$-trees.

The process we describe is a rewriting system based on the "divide and conquer" algorithm design paradigm (more precisely, "divide and marriage before conquest" technique, [**Bl99**]).

For a given generalised equation Ω_{v_0}, this branching process results in a locally finite and possibly infinite oriented rooted at v_0 tree T, $T = T(\Omega_{v_0})$. The vertices of the tree T are labelled by (constrained) generalised equations Ω_{v_i} over $\mathbb{F}$. The edges of the tree T are labelled by epimorphisms of the corresponding coordinate groups. Moreover, for every solution H of Ω_{v_0}, there exists a path in the tree T from the root vertex to a vertex v_l and a solution $H^{(l)}$ of Ω_{v_l} such that the solution H is a composition of the epimorphisms corresponding to the edges in the tree and the solution $H^{(l)}$. Conversely, every path from the root to a vertex v_l in T and any solution $H^{(l)}$ of Ω_{v_l} give rise to a solution of Ω_{v_0}, see Proposition 4.13.

The tree is constructed by induction on the height. Let v be a vertex of height n. One can check under the assumptions of which of the 15 cases described in Section 4.4 the generalised equation Ω_v falls. If Ω_v falls under the assumptions of Case 1 or Case 2, then v is a leaf of the tree T. Otherwise, using the combination of elementary and derived transformations (defined in Sections 4.2 and 4.3) given in the description of the corresponding (to v) case, one constructs finitely many generalised equations and epimorphisms from the coordinate group of Ω_v to the coordinate groups of the generalised equations constructed.

We finish this chapter (see Lemma 4.19) by proving that infinite branches of the tree T, as in the case of free groups, correspond to one of the following three cases:

(A) Case 7-10: Linear case (Levitt type, thin type);
(B) Case 12: Quadratic case (surface type, interval exchange type);
(C) Case 15: General case (toral type, axial type).

4.1. Preliminary definitions

In this section we give some definitions that we use throughout the text.

DEFINITION 4.1. Let Ω be a generalised equation. We partition the set $\Sigma = \Sigma(\Omega)$ of all closed sections of Ω into a disjoint union of two subsets
$$\Sigma(\Omega) = V\Sigma \cup C\Sigma.$$
The sections from $V\Sigma$ and $C\Sigma$, are called correspondingly, *variable*, and *constant* sections.

To organise the process properly, we partition the closed sections of Ω in another way into a disjoint union of two sets, which we refer to as *parts*:
$$\Sigma(\Omega) = A\Sigma \cup NA\Sigma$$
Sections from $A\Sigma$ are called *active*, and sections from $NA\Sigma$ are called *non-active*. We set
$$C\Sigma \subseteq NA\Sigma.$$
If not stated otherwise, we assume that all sections from $V\Sigma$ belong to the active part $A\Sigma$.

If $\sigma \in \Sigma$, then every item (or base) from σ is called active or non-active, depending on the type of σ.

DEFINITION 4.2. We say that a generalised equation Ω is in the *standard form* if the following conditions hold.
   (1) All non-active sections are located to the right of all active sections. Formally, there are numbers $1 \le \rho_A \le \rho_\Omega+1$ such that $[1, \rho_A]$ and $[\rho_A, \rho_\Omega+1]$ are, correspondingly, unions of all active and all non-active sections.
   (2) All constant bases belong to $C\Sigma$, and for every letter $a \in \mathcal{A}^{\pm 1}$ there is at most one constant base in Ω labelled by a.
   (3) Every free variable h_i of Ω belongs to a section from $C\Sigma$.

We will show in Lemma 4.12 that every generalised equation can be taken to the standard form.

4.2. Elementary transformations

In this section we describe *elementary transformations* of generalised equations. Recall that we consider only formally consistent generalised equations. In general, an elementary transformation ET associates to a generalised equation $\Omega = \langle \Upsilon, \Re_\Upsilon \rangle$ a finite set of generalised equations $\mathrm{ET}(\Omega) = \{\Omega_1, \ldots, \Omega_r\}$, $\Omega_i = \langle \Upsilon_i, \Re_{\Upsilon_i} \rangle$ and a collection of surjective homomorphisms $\theta_i : \mathbb{G}_{R(\Omega^*)} \to \mathbb{G}_{R(\Omega_i^*)}$ such that for every pair (Ω, H) there exists a unique pair $(\Omega_i, H^{(i)})$ such that the following diagram commutes.

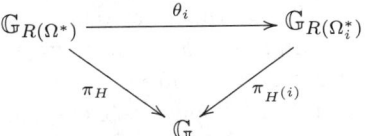

Since the pair $(\Omega_i, H^{(i)})$ is defined uniquely, we have a well-defined (partial) map $\mathrm{ET} : (\Omega, H) \to (\Omega_i, H^{(i)})$ (from and to the set of pairs (generalised equation, solution)).

Every elementary transformation is first described formally and then we give an example using graphic representations of generalised equations, the latter being much more intuitive. It is a good exercise to understand what how the elementary

4. THE PROCESS: CONSTRUCTION OF THE TREE T

transformations change the system of equations corresponding to the generalised equation.

ET 1: *Cutting a base*. Suppose that Ω contains a boundary connection (p, λ, q).

The transformation ET 1 carries Ω into a single generalised equation $\Omega_1 = \langle \Upsilon_1, \Re_{\Upsilon_1} \rangle$ which is obtained from Ω as follows. To obtain Υ_1 from Υ we
- replace (cut in p) the base λ by two new bases λ_1 and λ_2, and
- replace (cut in q) $\Delta(\lambda)$ by two new bases $\Delta(\lambda_1)$ and $\Delta(\lambda_2)$,

so that the following conditions hold.

If $\varepsilon(\lambda) = \varepsilon(\Delta(\lambda))$, then

$$\alpha(\lambda_1) = \alpha(\lambda),\ \beta(\lambda_1) = p,\quad \alpha(\lambda_2) = p,\ \beta(\lambda_2) = \beta(\lambda);$$
$$\alpha(\Delta(\lambda_1)) = \alpha(\Delta(\lambda)),\ \beta(\Delta(\lambda_1)) = q,\quad \alpha(\Delta(\lambda_2)) = q,\ \beta(\Delta(\lambda_2)) = \beta(\Delta(\lambda));$$

If $\varepsilon(\lambda) = -\varepsilon(\Delta(\lambda))$, then

$$\alpha(\lambda_1) = \alpha(\lambda),\ \beta(\lambda_1) = p,\quad \alpha(\lambda_2) = p,\ \beta(\lambda_2) = \beta(\lambda);$$
$$\alpha(\Delta(\lambda_1)) = q,\ \beta(\Delta(\lambda_1)) = \beta(\Delta(\lambda)),\quad \alpha(\Delta(\lambda_2)) = \alpha(\Delta(\lambda)),\ \beta(\Delta(\lambda_2)) = q;$$

Put

$$\varepsilon(\lambda_i) = \varepsilon(\lambda),\ \varepsilon(\Delta(\lambda_i)) = \varepsilon(\Delta(\lambda)), i = 1, 2.$$

Let (p', λ, q') be a boundary connection in Ω. If $p' < p$, then we replace (p', λ, q') by (p', λ_1, q'). If $p' > p$, then we replace (p', λ, q') by (p', λ_2, q'). Notice that from property (2) of Definition 3.9, it follows that (p', λ_1, q') (or (p', λ_2, q')) is a boundary connection in the new generalised equation.

We define the new generalised equation $\Omega_1 = \langle \Upsilon_1, \Re_{\Upsilon_1} \rangle$, by setting $\Re_{\Upsilon_1} = \Re_{\Upsilon}$. The resulting generalised equation Ω_1 is also formally consistent. Put $\text{ET}\,1(\Omega) = \{\Omega_1\}$, see Figure 1.

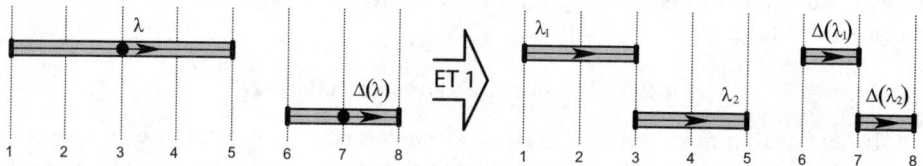

FIGURE 1. Elementary transformation ET 1: Cutting a base.

ET 2: *Transferring a base*. Let a base λ of a generalised equation Ω be contained in the base μ, i.e., $\alpha(\mu) \leq \alpha(\lambda) < \beta(\lambda) \leq \beta(\mu)$. Suppose that the boundaries $\alpha(\lambda)$ and $\beta(\lambda)$ are μ-tied, i.e. there are boundary connections of the form $(\alpha(\lambda), \mu, q_1)$ and $(\beta(\lambda), \mu, q_2)$. Suppose also that every λ-tied boundary is μ-tied.

The transformation ET 2 carries Ω into a single generalised equation $\Omega_1 = \langle \Upsilon_1, \Re_{\Upsilon_1} \rangle$ which is obtained from Ω as follows. To obtain Υ_1 from Υ we transfer λ from the base μ to the base $\Delta(\mu)$ and adjust all the basic and boundary equations (see Figure 2). Formally, we replace λ by a new base λ' such that $\alpha(\lambda') = q_1, \beta(\lambda') = q_2$ and replace each λ-boundary connection (p, λ, q) with a new one (p', λ', q) where p and p' are related by a μ-boundary connection (p, μ, p').

By definition, set $\Re_{\Upsilon_1} = \Re_{\Upsilon}$. We therefore defined a generalised equation $\Omega = \langle \Upsilon_1, \Re_{\Upsilon_1} \rangle$, set $\text{ET}\,2(\Omega) = \{\Omega_1\}$.

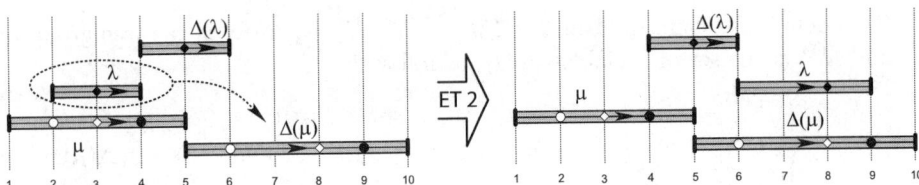

FIGURE 2. Elementary transformation ET 2: Transferring a base.

ET 3: *Removing a pair of matched bases.* Let μ and $\Delta(\mu)$ be a pair of matched bases in Ω. Since Ω is formally consistent, one has $\varepsilon(\mu) = \varepsilon(\Delta(\mu))$, $\beta(\mu) = \beta(\Delta(\mu))$ and every μ-boundary connection is of the form (p, μ, p).

The transformation ET 3 applied to Ω results in a single generalised equation $\Omega_1 = \langle \Upsilon_1, \Re_{\Upsilon_1} \rangle$ which is obtained from Ω by removing the pair of bases $\mu, \Delta(\mu)$ with all the μ-boundary connections and setting $\Re_{\Upsilon_1} = \Re_\Upsilon$, see Figure 3.

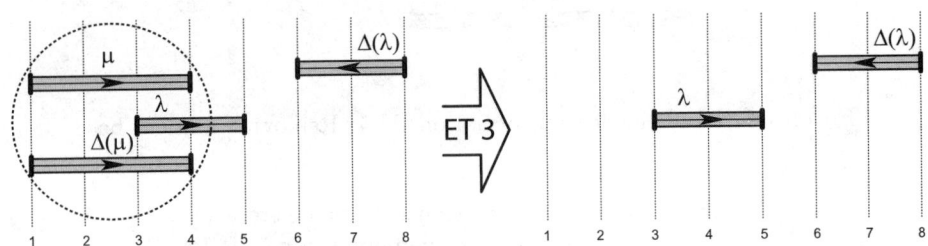

FIGURE 3. Elementary transformation ET 3: Removing a pair of matched bases.

REMARK 4.3. Observe that for $i = 1, 2, 3$, the set $\mathrm{ET}\,i(\Omega)$ consists of a single generalised equation Ω_1, such that Ω and Ω_1 have the same set of variables h and $\Re_\Upsilon = \Re_{\Upsilon_1}$. The identity isomorphism $\tilde{\theta}_1 : F[h] \to F[h]$, where F is the free group on $\mathcal{A}$, trivially induces a $\mathbb{G}$-isomorphism θ_1 from $\mathbb{G}_{R(\Omega^*)}$ to $\mathbb{G}_{R(\Omega_1^*)}$ and an F-isomorphism θ_1' from $F_{R(\Upsilon^*)}$ to $F_{R(\Upsilon_1^*)}$.

Moreover, if H is a solution of Ω, then the tuple H is a solution of Ω_1, since the substitution of H into the equations of Ω_1 result in graphical equalities.

ET 4: *Removing a linear base.* Suppose that in Ω a variable base μ does not intersect any other variable base, i.e. the items $h_{\alpha(\mu)}, \ldots, h_{\beta(\mu)-1}$ are contained only in one variable base μ. Moreover, suppose that all boundaries that intersect μ are μ-tied, i.e. for every i, $\alpha(\mu) < i < \beta(\mu)$ there exists a boundary $\mathfrak{t}(i)$ such that $(i, \mu, \mathfrak{t}(i))$ is a boundary connection in Ω. Since Ω is formally consistent, we have $\mathfrak{t}(\alpha(\mu)) = \alpha(\Delta(\mu))$ and $\mathfrak{t}(\beta(\mu)) = \beta(\Delta(\mu))$ if $\varepsilon(\mu)\varepsilon(\Delta(\mu)) = 1$, and $\mathfrak{t}(\alpha(\mu)) = \beta(\Delta(\mu))$ and $\mathfrak{t}(\beta(\mu)) = \alpha(\Delta(\mu))$ if $\varepsilon(\mu)\varepsilon(\Delta(\mu)) = -1$.

The transformation ET 4 carries Ω into a single generalised equation $\Omega_1 = \langle \Upsilon_1, \Re_{\Upsilon_1} \rangle$ which is obtained from Ω by deleting the pair of bases μ and $\Delta(\mu)$; deleting all the boundaries $\alpha(\mu) + 1, \ldots, \beta(\mu) - 1$, deleting all μ-boundary connections, re-enumerating the remaining boundaries and setting $\Re_{\Upsilon_1} = \Re_\Upsilon$.

54   4. THE PROCESS: CONSTRUCTION OF THE TREE T

We define the epimorphism $\tilde{\theta}_1 : F[h] \to F[h^{(1)}]$, where F is the free group on $\mathcal{A}$ and $h^{(1)}$ is the set of variables of Ω_1, as follows:

$$\tilde{\theta}_1(h_j) = \begin{cases} h_j^{(1)}, & \text{if } j < \alpha(\mu) \text{ or } j \geq \beta(\mu); \\ h_{t(j)}^{(1)} \ldots h_{t(j+1)-1}^{(1)}, & \text{if } \alpha + 1 \leq j \leq \beta(\mu) - 1 \text{ and } \varepsilon(\mu) = \varepsilon(\Delta(\mu)); \\ h_{t(j-1)}^{(1)} \ldots h_{t(j+1)}^{(1)}, & \text{if } \alpha + 1 \leq j \leq \beta(\mu) - 1 \text{ and } \varepsilon(\mu) = -\varepsilon(\Delta(\mu)). \end{cases}$$

It is not hard to see that $\tilde{\theta}_1$ induces a $\mathbb{G}$-isomorphism $\theta_1 : \mathbb{G}_{R(\Omega^*)} \to \mathbb{G}_{R(\Omega_1^*)}$ and an F-isomorphism θ_1' from $F_{R(\Upsilon^*)}$ to $F_{R(\Upsilon_1^*)}$. Furthermore, if H is a solution of Ω the tuple $H^{(1)}$ (obtained from H in the same way as $h^{(1)}$ is obtained from h) is a solution of Ω_1.

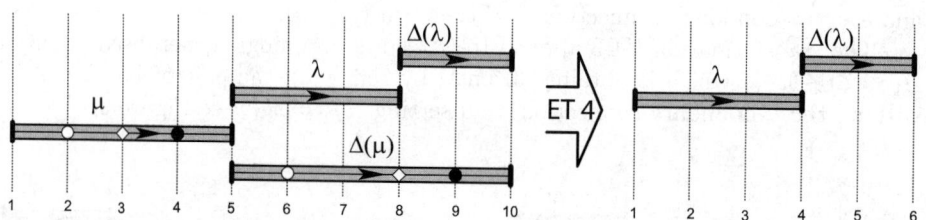

FIGURE 4. Elementary transformation ET 4: Removing a linear base.

REMARK 4.4. Every time when we transport or delete a closed section $[i, j]$ (see ET 4, D 2, D 5, D 6), we re-enumerate the boundaries as follows.

The re-enumeration of the boundaries is defined by the correspondence $\mathcal{C} : \mathcal{BD}(\Omega) \to \mathcal{BD}(\Omega_1)$, $\Omega_1 \in \text{ET 4}(\Omega)$:

$$\mathcal{C} : k \mapsto \begin{cases} k, & \text{if } k \leq i; \\ k - (j - i), & \text{if } k \geq j. \end{cases}$$

Naturally, in this case we write $\Re_{\Upsilon_1} = \Re_\Upsilon$ meaning that

$$\Re_{\Omega_1}(h_k) = \begin{cases} \Re_\Upsilon(h_k), & \text{if } k \leq i; \\ \Re_\Upsilon(h_{k-(j-i)}), & \text{if } k \geq j. \end{cases}$$

ET 5: *Introducing a new boundary.* Suppose that a boundary p intersects a base μ and p is not μ-tied.

The transformation ET 5 μ-ties the boundary p in all possible ways, producing finitely many different generalised equations. To this end, let q be a boundary intersecting $\Delta(\mu)$. Then we perform one of the following two transformations (see Figure 5):

(1) Introduce the boundary connection (p, μ, q) in Υ, provided that the resulting generalised equation Υ_q is formally consistent. Define the set $\Re_{\Upsilon_q} \subseteq h \times h$ to be the minimal subset that contains $\Re_\Upsilon$, is symmetric and satisfies condition $(\star)$, see Definition 3.4. This defines the generalised equation $\Omega_q = \langle \Upsilon_q, \Re_{\Upsilon_q} \rangle$.

The identity isomorphism $\tilde{\theta}_q : F[h] \to F[h]$, where F is the free group on $\mathcal{A}$, trivially induces a $\mathbb{G}$-epimorphism θ_q from $\mathbb{G}_{R(\Omega^*)}$ to $\mathbb{G}_{R(\Omega_q^*)}$ and an F-epimorphism θ_q' from $F_{R(\Upsilon^*)}$ to $F_{R(\Upsilon_q^*)}$.

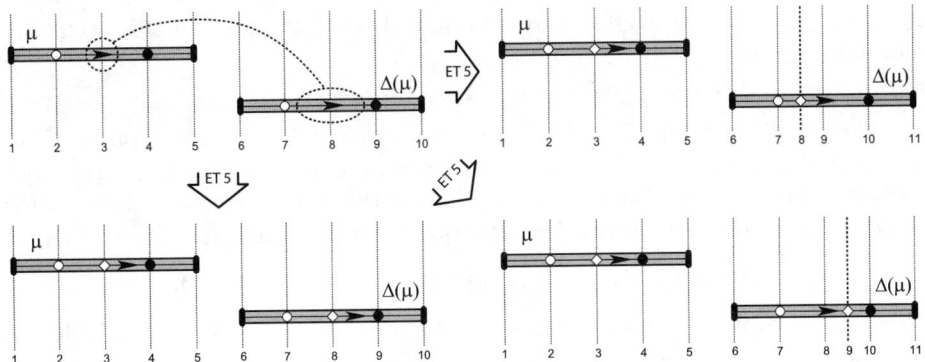

FIGURE 5. Elementary transformation ET 5: Introducing a new boundary.

Observe that θ_q is not necessarily an isomorphism. More precisely, θ_q is not an isomorphism whenever the boundary equation corresponding to the boundary connection (p, μ, q) does not belong to $R(\Omega^*)$. In this case θ_q is a proper epimorphism. Moreover, if H is a solution of Ω, then the tuple H is a solution of Ω_q.

(2) Introduce a new boundary q' between q and $q+1$; introduce a new boundary connection (p, μ, q') in Υ. Denote the resulting generalised equation by $\Upsilon_{q'}$. Set (below we assume that the boundaries of $\Upsilon_{q'}$ are labelled $1, 2, \ldots, q, q', q+1, \ldots, \rho_{\Upsilon_{q'}} + 1$):

$$\Re_{\Upsilon_{q'}}(h_i) = \begin{cases} \Re_{\Upsilon}(h_q), & \text{if } i = q, q'; \\ \Re_{\Upsilon}(h_i), & \text{otherwise.} \end{cases}$$

We then make the set $\Re_{\Upsilon_{q'}}$ symmetric and complete it in such a way that it satisfies condition $(\star)$, see Definition 3.4. This defines the generalised equation $\Omega_{q'} = \langle \Upsilon_{q'}, \Re_{\Upsilon_{q'}} \rangle$.

We define the F-monomorphism $\tilde{\theta}_{q'} : F[h] \hookrightarrow F[h^{(q')}]$, where F is the free group on $\mathcal{A}$ and $h^{(q')}$ is the set of variables of $\Omega_{q'}$, as follows:

$$\tilde{\theta}(h_i) = \begin{cases} h_i, & \text{if } i \neq q; \\ h_{q'-1} h_{q'}, & \text{if } i = q. \end{cases}$$

Observe that the F-monomorphism $\tilde{\theta}_{q'}$ induces a $\mathbb{G}$-isomorphism $\theta_{q'} : \mathbb{G}_{R(\Omega^*)} \to \mathbb{G}_{R(\Omega_{q'}^*)}$ and an F-isomorphism $\theta'_{q'}$ from $F_{R(\Upsilon^*)}$ to $F_{R(\Upsilon_{q'}^*)}$. Moreover, if H is a solution of Ω, then the tuple $H^{(q')}$ (obtained from H in the same way as $h^{h^{(q')}}$ is obtained from h) is a solution of $\Omega_{q'}$.

LEMMA 4.5. *Let $\Omega_1 \in \{\Omega_i\} = \mathrm{ET}(\Omega)$ be a generalised equation obtained from Ω by an elementary transformation ET and let $\theta_1 : \mathbb{G}_{R(\Omega^*)} \to \mathbb{G}_{R(\Omega_1^*)}$ be the corresponding epimorphism. There exists an algorithm which determines whether or not the epimorphism θ_1 is a proper epimorphism.*

PROOF. The only non-trivial case is when $\mathrm{ET} = \mathrm{ET}\,5$ and no new boundaries were introduced. In this case Ω_1 is obtained from Ω by adding a new boundary

equation $s = 1$, which is effectively determined by Ω and Ω_1. In this event, the coordinate group
$$\mathbb{G}_{R(\Omega_1{}^*)} = \mathbb{G}_{R(\{\Omega \cup \{s\}\}^*)}$$
is a quotient of the group $\mathbb{G}_{R(\Omega^*)}$. The homomorphism θ_1 is an isomorphism if and only if $R(\Omega^*) = R(\{\Omega \cup \{s\}\}^*)$, or, equivalently, $s \in R(\Omega^*)$. The latter condition holds if and only if s vanishes on all solutions of the system of equations $\Omega^* = 1$ in $\mathbb{G}$, i.e. if the following universal formula (quasi identity) holds in $\mathbb{G}$:
$$\forall x_1 \ldots \forall x_\rho (\Omega^*(x_1, \ldots, x_\rho) = 1 \to s(x_1, \ldots, x_\rho) = 1).$$
This can be verified effectively, since the universal theory of $\mathbb{G}$ is decidable, see [**DL04**]. □

LEMMA 4.6. *Let $\Omega_1 \in \{\Omega_i\} = \mathrm{ET}(\Omega)$ be a generalised equation obtained from Ω by an elementary transformation ET and let $\theta_1 : \mathbb{G}_{R(\Omega^*)} \to \mathbb{G}_{R(\Omega_1^*)}$ be the corresponding epimorphism. Then there exists a homomorphism*
$$\tilde{\theta}_1 : F[h_1, \ldots, h_{\rho_\Omega}] \to F[h_1, \ldots, h_{\rho_{\Omega_1}}]$$
such that $\tilde{\theta}_1$ induces an epimorphism
$$\theta_1' : F_{R(\Upsilon^*)} \to F_{R(\Upsilon_1^*)}$$
and the epimorphism θ_1. In other words, the following diagram commutes:

$$\begin{array}{ccccc}
\mathbb{G}_{R(\Omega^*)} & \longleftarrow & F[h_1, \ldots, h_{\rho_\Omega}] & \longrightarrow & F_{R(\Upsilon^*)} \\
{\scriptstyle \theta_1} \downarrow & & {\scriptstyle \tilde{\theta}} \downarrow & & \downarrow {\scriptstyle \theta'} \\
\mathbb{G}_{R(\Omega_1^*)} & \longleftarrow & F[h_1, \ldots, h_{\rho_{\Omega_1}}] & \longrightarrow & F_{R(\Upsilon_1^*)}
\end{array}$$

PROOF. Follows by examining the definition of θ_1 for every elementary transformation ET 1 – ET 5. □

4.3. Derived transformations

In this section we describe several useful transformations of generalised equations. Some of them are finite sequences of elementary transformations, others result in equivalent generalised equations but cannot be presented as a composition of finitely many elementary transformations.

In general, a *derived transformation* D associates to a generalised equation $\Omega = \langle \Upsilon, \Re_\Upsilon \rangle$ a finite set of generalised equations $D(\Omega) = \{\Omega_1, \ldots, \Omega_r\}$, $\Omega_i = \langle \Upsilon_i, \Re_{\Upsilon_i} \rangle$ and a collection of surjective homomorphisms $\theta_i : \mathbb{G}_{R(\Omega^*)} \to \mathbb{G}_{R(\Omega_i^*)}$ such that for every pair (Ω, H) there exists a unique pair $(\Omega_i, H^{(i)})$ such that the following diagram commutes.

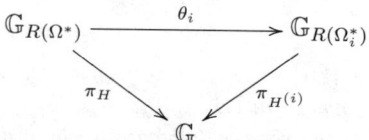

Since the pair $(\Omega_i, H^{(i)})$ is defined uniquely, we have a well-defined map $D : (\Omega, H) \to (\Omega_i, H^{(i)})$.

4.3. DERIVED TRANSFORMATIONS

D 1 : *Closing a section.* Let $\sigma = [i, j]$ be a section of Ω. The transformation D 1 makes the section σ closed. To perform D 1, using ET 5, we μ-tie the boundary i (the boundary j) in every base μ containing i (j, respectively). Using ET 1, we cut all the bases containing i (or j) in the boundary i (or in j), see Figure 6.

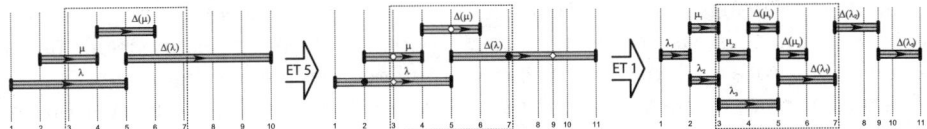

FIGURE 6. Derived transformation D 1: Closing a section.

D 2 : *Transporting a closed section.* Let σ be a closed section of a generalised equation Ω. The derived transformation D 2 takes Ω to a single generalised equation Ω_1 obtained from Ω by cutting σ out from the interval $[1, \rho_\Omega + 1]$ together with all the bases, and boundary connections on σ and moving σ to the end of the interval or between two consecutive closed sections of Ω, see Figure 7. Then we re-enumerate all the items and boundaries of the generalised equation obtained as appropriate, see Remark 4.4. Clearly, the original equation Ω and the new one Ω_1 have the same solution sets and their coordinate groups are isomorphic (the isomorphism is induced by a permutation of the variables h).

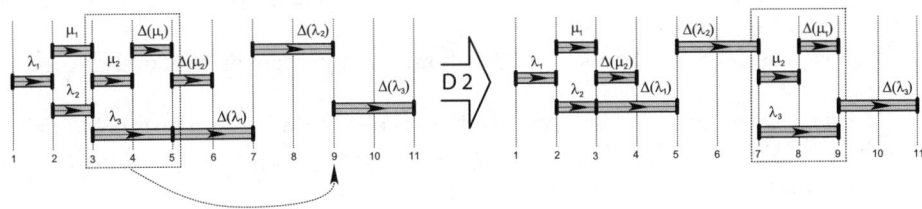

FIGURE 7. Derived transformation D 2: Transporting a closed section.

D 3 : *Completing a cut.* Let Ω be constrained a generalised equation. The derived transformation D 3 carries Ω into a single generalised equation $\widetilde{\Omega} = \Omega_1$ by applying ET 1 to all boundary connections in Ω. Formally, for every boundary connection (p, μ, q) in Ω we cut the base μ in p applying ET 1. Clearly, the generalised equation $\widetilde{\Omega}$ does not depend on the choice of a sequence of transformations ET 1. Since, by Remark 4.3, ET 1 induces the identity isomorphism between the coordinate groups, equations Ω and $\widetilde{\Omega}$ have isomorphic coordinate groups and the isomorphism is induced by the identity map $\mathbb{G}[h] \to \mathbb{G}[h]$.

D 4 : *Kernel of a generalised equation.* The derived transformation D 4 applied to a generalised equation Ω results in a single generalised equation $\text{Ker}(\Omega) = \text{D } 4(\Omega)$ constructed below.

Applying D 3, if necessary, one can assume that a generalised equation $\Omega = \langle \Upsilon, \Re_\Upsilon \rangle$ does not contain boundary connections. An active base $\mu \in A\Sigma(\Upsilon)$ is called *eliminable* if at least one of the following holds:

a) μ contains an item h_i with $\gamma(h_i) = 1$;

b) at least one of the boundaries $\alpha(\mu), \beta(\mu)$ is different from $1, \rho+1$ and it does not touch any other base (except μ).

An *elimination process* for Υ consists of consecutive removals (*eliminations*) of eliminable bases until there are no eliminable bases left in the generalised equation. The resulting generalised equation is called the *kernel* of Υ and we denote it by $\mathrm{Ker}(\Upsilon)$. It is easy to see that $\mathrm{Ker}(\Upsilon)$ does not depend on the choice of the elimination process. Indeed, if Υ has two different eliminable bases μ_1, μ_2, and elimination of μ_i results in a generalised equation Υ_i then by induction on the number of eliminations $\mathrm{Ker}(\Upsilon_i)$ is defined uniquely for $i = 1, 2$. Obviously, μ_1 is still eliminable in Υ_2, as well as μ_2 is eliminable in Υ_1. Now eliminating μ_1 and μ_2 from Υ_2 and Υ_1 respectively, we get the same generalised equation Υ_0. By induction $\mathrm{Ker}(\Upsilon_1) = \mathrm{Ker}(\Upsilon_0) = \mathrm{Ker}(\Upsilon_2)$ hence the result.

We say that a variable h_i *belongs to the kernel*, if h_i either belongs to at least one base in the kernel, or is constant.

For a generalised equation Ω we denote by $\overline{\Omega} = \langle \overline{\Upsilon}, \Re_{\overline{\Upsilon}} \rangle$ the generalised equation which is obtained from Ω by deleting all free variables in Υ and setting $\Re_{\overline{\Upsilon}}$ to be the restriction of $\Re_{\Upsilon}$ to the set of variables in $\overline{\Upsilon}$. Obviously,

$$\mathbb{G}_{R(\Omega^*)} = \mathbb{G}[h_1, \ldots, h_{\rho_{\overline{\Omega}}}, y_1, \ldots, y_k] \Big/ R(\overline{\Omega}^* \cup \{[y_i, h_j] \mid \Re_{\Upsilon}(y_i, h_j)\}),$$

where $\{y_1, \ldots y_k\}$ is the set of free variables in Υ and h_j is a variable of Ω (in particular, h_j may be one of the free variables $y_1, \ldots, y_k$).

Let us consider what happens in the elimination process on the level of coordinate groups.

We start with the case when just one base is eliminated. Let μ be an eliminable base in $\Upsilon = \Upsilon(h_1, \ldots, h_\rho)$. Denote by Υ_1 the generalised equation obtained from Υ by eliminating μ and let $\Omega_1 = \langle \Upsilon_1, \Re_{\Upsilon_1} \rangle$, where $\Re_{\Upsilon_1} = \Re_{\Upsilon}$.

(1) Suppose $h_i \in \mu$ and $\gamma(h_i) = 1$, i.e. μ falls under the assumption a) of the definition of an eliminable base. Then the variable h_i occurs only once in Υ (in the equation $h(\mu)^{\varepsilon(\mu)} = h(\Delta(\mu))^{\varepsilon(\Delta(\mu))}$ corresponding to the base μ, denote this equation by s_μ). Therefore, in the coordinate group $\mathbb{G}_{R(\Omega^*)}$ the relation s_μ can be written as $h_i = w_\mu$, where w_μ does not contain h_i. Consider the generalised equation Υ' obtained from Υ by deleting the equation s_μ and the item h_i. The presentation of the coordinate group of Υ' is obtained from the presentation of the coordinate group of Υ using a sequence of Tietze transformations, thus these groups are isomorphic. We define the relation $\Re_{\Upsilon'}$ as the restriction of $\Re_{\Upsilon}$ to the set $h \setminus \{h_i\}$, and $\Omega' = \langle \Upsilon', \Re_{\Upsilon'} \rangle$.

It follows that

$$\mathbb{G}_{R(\Omega^*)} \simeq \mathbb{G}[h_1, \ldots, h_{i-1}, h_{i+1}, \ldots, h_\rho] \Big/ R(\Omega'^* \cup \{[h_j, w_\mu] \mid \Re_{\Upsilon}(h_j, h_i)\}).$$

We therefore get that

$$\mathbb{G}_{R(\Omega_1^*)} \simeq \mathbb{G}[h_1, \ldots, h_\rho] \Big/ R(\Omega'^* \cup \{[h_i, h_j] \mid \Re_{\Upsilon}(h_i, h_j)\}),$$

and

(4.1)
$$\mathbb{G}_{R(\Omega^*)} \simeq \mathbb{G}[h_1,\ldots,h_{i-1},h_{i+1},\ldots,h_\rho]\big/R(\Omega'^* \cup \{[h_j,w_\mu] \mid \Re_\Upsilon(h_j,h_i)\})$$

$$\simeq \mathbb{G}[h_1,\ldots,h_{\rho_{\overline{\Omega}_1}},z_1,\ldots,z_l]\bigg/R\begin{pmatrix} \overline{\Omega}_1^* \cup \{[h_j,w_\mu] \mid \Re_\Upsilon(h_j,h_i)\} \\ \cup\{[z_k,h_j] \mid \Re_\Upsilon(z_k,h_j)\} \\ \cup\{[z_k,w_\mu] \mid \Re_\Upsilon(z_k,h_i)\} \end{pmatrix},$$

where $\{z_1,\ldots,z_l\}$ are the free variables of Ω_1 distinct from h_i. Note that all the groups and equations which occur above can be found effectively.

(2) Suppose now that μ falls under the assumptions of case b) of the definition of an eliminable base, with respect to a boundary i. Then in all the equations of the generalised equation Ω but s_μ, the variable h_{i-1} occurs as a part of the subword $(h_{i-1}h_i)^{\pm 1}$. In the equation s_μ the variable h_{i-1} either occurs once or it occurs precisely twice and in this event the second occurrence of h_{i-1} (in $\Delta(\mu)$) is a part of the subword $(h_{i-1}h_i)^{\pm 1}$. It is obvious that the tuple

$$(h_1,\ldots,h_{i-2},h_{i-1},h_{i-1}h_i,h_{i+1},\ldots,h_\rho) = (h_1,\ldots,h_{i-2},h_{i-1},h'_i,h_{i+1},\ldots,h_\rho)$$

forms a basis of the ambient free group generated by $(h_1,\ldots,h_\rho)$. We further assume that Ω is a system of equations in these variables. Notice, that in the new basis, the variable h_{i-1} occurs only once in the generalised equation Ω, in the equation s_μ. Hence, as in case (1), the relation s_μ can be written as $h_{i-1} = w_\mu$, where w_μ does not contain h_{i-1}.

Therefore, if we eliminate the relation s_μ, then the set $Y = (h_1,\ldots,h_{i-2},h'_i,h_{i+1},\ldots,h_\rho)$ is a generating set of $\mathbb{G}_{R(\Omega^*)}$. In the coordinate group $\mathbb{G}_{R(\Omega^*)}$ the commutativity relations $\{[h_j,h_k] \mid \Re_\Upsilon(h_j,h_k)\}$ rewrite as follows:

$$\left\{\begin{array}{ll} [h_j,h_k], & \text{if } \Re_\Upsilon(h_j,h_k) \text{ and } j,k \notin \{i,i-1\}; \\ [w_\mu,h_j], & \text{if } \Re_\Upsilon(h_{i-1},h_j); \\ [w_\mu^{-1}h'_i,h_j], & \text{if } \Re_\Upsilon(h_i,h_j) \end{array}\right\}.$$

This shows that

$$\mathbb{G}_{R(\Omega^*)} \simeq \mathbb{G}[h_1,\ldots,h_{i-2},h'_i,h_{i+1},\ldots,h_\rho]\bigg/R\begin{pmatrix} \Omega'^* \cup \{[w_\mu,h_j] \mid \Re_\Upsilon(h_{i-1},h_j)\} \\ \cup\{[w_\mu^{-1}h'_i,h_j] \mid \Re_\Upsilon(h_i,h_j)\} \end{pmatrix},$$

where $\Omega' = \langle \Upsilon',\Re_{\Upsilon'} \rangle$ is the generalised equation obtained from Ω_1 by deleting the boundary i and $\Re_{\Upsilon'}$ is the restriction of $\Re_{\Upsilon_1}$ to the set $h \setminus \{h_{i-1},h_i\}$. Denote by Ω'' the generalised equation obtained from Ω' by adding a free variable z. It now follows that

$$\mathbb{G}_{R(\Omega_1^*)} \simeq \mathbb{G}[h_1,\ldots,h_{i-2},h_{i+1},\ldots,h_\rho,z]\bigg/R\begin{pmatrix} \Omega''^* \cup \{[z,h_j] \mid \Re_{\Upsilon_1}(h_{i-1},h_j)\} \\ \cup\{[z^{-1}h'_i,h_j] \mid \Re_{\Upsilon_1}(h_i,h_j)\} \end{pmatrix}$$

$$\simeq \mathbb{G}[h_1,\ldots,h_{i-2},h_{i+1},\ldots,h_\rho,z]\bigg/R\begin{pmatrix} \Omega'^* \cup \{[z,h_j] \mid \Re_{\Upsilon_1}(h_{i-1},h_j)\} \\ \cup\{[z^{-1}h'_i,h_j] \mid \Re_{\Upsilon_1}(h_i,h_j)\} \end{pmatrix}$$

and that
(4.2)
$$\mathbb{G}_{R(\Omega^*)} \simeq \mathbb{G}[h_1,\ldots,h_{i-2},h'_i,h_{i+1},\ldots,h_\rho] \Big/ R \left(\begin{array}{c} \Omega'^* \cup \{[w_\mu, h_j] \mid \Re_\Upsilon(h_{i-1}, h_j)\} \\ \cup \{[w_\mu^{-1} h'_i, h_j] \mid \Re_\Upsilon(h_i, h_j)\} \end{array} \right)$$

$$\simeq \mathbb{G}[h_1,\ldots,h_{\rho_{\overline{\Omega''}}}, z_1,\ldots,z_l] \Big/ R \left(\begin{array}{c} \overline{\Omega''}^* \cup \{[w_\mu^{-1} h'_i, h_j] \mid \Re_{\Upsilon_1}(h_i, h_j)\} \\ \cup \{[w_\mu^{-1} h'_i, z_j] \mid \Re_{\Upsilon_1}(h_i, z_j)\} \\ \cup \{[w_\mu, h_j] \mid \Re_{\Upsilon_1}(h_{i-1}, h_j)\} \\ \cup \{[w_\mu, z_j] \mid \Re_{\Upsilon_1}(h_{i-1}, z_j)\} \\ \cup \{[z_j, h_k] \mid \Re_{\Upsilon'}(z_j, h_k)\} \end{array} \right)$$

where $\{z_1, \ldots, z_l\}$ are the free variables of Ω'' distinct from z. Notice that all the groups and equations which occur above can be found effectively.

By induction on the number of steps in the elimination process we obtain the following lemma.

LEMMA 4.7. *In the above notation,*
$$\mathbb{G}_{R(\Omega^*)} \simeq \mathbb{G}[h_1,\ldots,h_{\rho_{\overline{\mathrm{Ker}(\Omega)}}}, z_1,\ldots,z_l] \Big/ R(\overline{\mathrm{Ker}(\Omega)}^* \cup \mathcal{K}),$$
where $\{z_1, \ldots, z_l\}$ is a set of free variables of $\mathrm{Ker}(\Omega)$, and $\mathcal{K}$ is a certain computable set of commutators of words in $h \cup Z$.

PROOF. Let
$$\Omega = \Omega_0 \to \Omega_1 \to \ldots \to \Omega_k = \mathrm{Ker}(\Omega),$$
where Ω_{j+1} is obtained from Ω_j by eliminating an eliminable base, $j = 0, \ldots, k-1$. It is easy to see (by induction on k) that for every $j = 0, \ldots, k-1$
$$\overline{\mathrm{Ker}(\Omega_j)} = \overline{\mathrm{Ker}(\overline{\Omega_j})}.$$
Moreover, if Ω_{j+1} is obtained from Ω_j as in case (2) above, then (in the above notation)
$$\overline{(\mathrm{Ker}(\Omega_j))_1} = \overline{\mathrm{Ker}(\Omega''_j)}.$$
The statement of the lemma now follows from the remarks above and Equations (4.1) and (4.2). Note that the cardinality l of the set $\{z_1, \ldots, z_l\}$ is the number of free variables of $\mathrm{Ker}(\Omega)$ minus k. □

D 5 : *Entire transformation.* In order to define the derived transformation D 5 we need to introduce several notions. A base μ of the generalised equation Ω is called a *leading* base if $\alpha(\mu) = 1$. A leading base is called a *carrier* if for any other leading base λ we have $\beta(\lambda) \leq \beta(\mu)$. Let μ be a carrier base of Ω. Any active base $\lambda \neq \mu$ with $\beta(\lambda) \leq \beta(\mu)$ is called a *transfer* base (with respect to μ).

Suppose now that Ω is a generalised equation with $\gamma(h_i) \geq 2$ for each h_i in the active part of Ω. The *entire transformation* is the following sequence of elementary transformations.

We fix a carrier base μ of Ω. For any transfer base λ we μ-tie (applying ET 5) all boundaries that intersect λ. Using ET 2 we transfer all transfer bases from μ onto $\Delta(\mu)$. Now, there exists some $k < \beta(\mu)$ such that $h_1, \ldots, h_k$ belong to only one base μ, while h_{k+1} belongs to at least two bases. Applying ET 1 we cut μ in the boundary $k+1$. Finally, applying ET 4, we delete the section $[1, k+1]$, see Figure 12.

4.3. DERIVED TRANSFORMATIONS

Let ρ_A be the boundary between the active and non-active parts of Ω, i.e. $[1,\rho_A]$ is the active part $A\Sigma$ of Ω and $[\rho_A, \rho+1]$ is the non-active part $NA\Sigma$ of Ω.

DEFINITION 4.8. For a pair (Ω, H), we introduce the following notation

(4.3) $$d_{A\Sigma}(H) = \sum_{i=1}^{\rho_A - 1} |H_i|,$$

(4.4) $$\psi_{A\Sigma}(H) = \sum_{\mu \in \omega_1} |H(\mu)| - 2d_{A\Sigma}(H),$$

where ω_1 is the set of all variable bases ν for which either ν or $\Delta(\nu)$ belongs to the active part $[1, \rho_A]$ of Ω.

We call the number $\psi_{A\Sigma}(H)$ the *excess* of the solution H of the generalised equation Ω.

Every item h_i of the section $[1, \rho_A]$ belongs to at least two bases, each of these bases belongs to $A\Sigma$, hence $\psi_{A\Sigma}(H) \geq 0$.

Notice, that if for every item h_i of the section $[1, \rho_A]$ one has $\gamma(h_i) = 2$, for every solution H of Ω the excess $\psi_{A\Sigma}(H) = 0$. Informally, in some sense, the excess of H measures how far the generalised equation Ω is from being quadratic (every item in the active part is covered twice).

LEMMA 4.9. *Let Ω be a generalised equation such that every item h_i from the active part $A\Sigma = [1, \rho_A]$ of Ω is covered at least twice. Suppose that $D\,5(\Omega, H) = (\Omega_i, H^{(i)})$ and the carrier base μ of Ω and its dual $\Delta(\mu)$ belong to the active part of Ω. Then the excess of the solution H equals the excess of the solution $H^{(i)}$, i.e. $\psi_{A\Sigma}(H) = \psi_{A\Sigma}(H^{(i)})$.*

PROOF. By the definition of the entire transformation $D\,5$, it follows that the word $H^{(i)}[1, \rho_{\Omega_i} + 1]$ is a terminal subword of the word $H[1, \rho+1]$, i.e.

$$H[1, \rho+1] \doteq U_i H^{(i)}[1, \rho_{\Omega_i} + 1], \text{ where } \rho = \rho_\Omega.$$

On the other hand, since $\mu, \Delta(\mu) \in A\Sigma$, the non-active parts of Ω and Ω_i coincide. Therefore, $H[\rho_A, \rho+1]$ is the terminal subword of the word $H^{(i)}[1, \rho_{\Omega_i}+1]$, i.e. the following graphical equality holds:

$$H^{(i)}[1, \rho_{\Omega_i} + 1] \doteq V_i H[\rho_A, \rho+1].$$

So we have

(4.5) $$d_{A\Sigma}(H) - d_{A\Sigma}(H^{(i)}) = |H[1, \rho_A]| - |V_i| = |U_i| = |H(\mu)| - |H^{(i)}(\mu)|,$$

where, in the above notation, if $k = \beta(\mu) - 1$ (and thus μ has been completely eliminated in $\Omega^{(i)}$), then $H^{(i)}(\mu) = 1$; otherwise, $H^{(i)}(\mu) = H[k+1, \beta(\mu)]$.

From (4.4) and (4.5) it follows that $\psi_{A\Sigma}(H) = \psi_{A\Sigma}(H^{(i)})$. □

D6 : *Identifying closed constant sections.* Let λ and μ be two constant bases in Ω with labels a^{ϵ_λ} and a^{ϵ_μ}, where $a \in \mathcal{A}^{\pm 1}$ and $\epsilon_\lambda = \epsilon_\mu$, $\epsilon_\mu \in \{1, -1\}$. Suppose that the sections $\sigma(\lambda) = [i, i+1]$ and $\sigma(\mu) = [j, j+1]$ are closed.

The transformation D6 applied to Ω results in a single generalised equation Ω_1 which is obtained from Ω in the following way, see Figure 8. Introduce a new variable base η with its dual $\Delta(\eta)$ such that

$$\sigma(\eta) = [i, i+1], \ \sigma(\Delta(\eta)) = [j, j+1], \ \varepsilon(\eta) = \epsilon_\lambda, \ \varepsilon(\Delta(\eta)) = \epsilon_\mu.$$

Then we transfer all bases from η onto $\Delta(\eta)$ using ET2, remove the bases η and $\Delta(\eta)$, remove the item h_i, re-enumerate the remaining items and adjust the relation $\Re_\Upsilon$ as appropriate, see Remark 4.4.

The corresponding homomorphism $\theta_1 : \mathbb{G}_{R(\Omega^*)} \to \mathbb{G}_{R(\Omega_1^*)}$ is induced by the composition of the homomorphisms defined by the respective elementary transformations. Obviously, θ_1 is an isomorphism.

LEMMA 4.10. *Let Ω_i be a generalised equation and $H^{(i)}$ be a solution of Ω_i so that*

$$\text{ET} : (\Omega, H) \to (\Omega_i, H^{(i)}) \ or \ \text{D} : (\Omega, H) \to (\Omega_i, H^{(i)}).$$

Then one has $|H^{(i)}| \leq |H|$. Furthermore, in the case that ET = ET4 *or* D = D5 *this inequality is strict $|H^{(i)}| < |H|$.*

PROOF. Proof is by straightforward examination of descriptions of elementary and derived transformations. □

LEMMA 4.11. *Let $\Omega_1 \in \{\Omega_i\} = \text{ET}(\Omega)$ be a generalised equation obtained from Ω by a derived transformation D and let $\theta_1 : \mathbb{G}_{R(\Omega^*)} \to \mathbb{G}_{R(\Omega_1^*)}$ be the corresponding epimorphism. Then there exists a homomorphism*

$$\tilde{\theta}_1 : F[h_1, \ldots, h_{\rho_\Omega}] \to F[h_1, \ldots, h_{\rho_{\Omega_1}}]$$

such that $\tilde{\theta}_1$ induces an epimorphism

$$\theta_1' : F_{R(\Upsilon^*)} \to F_{R(\Upsilon_1^*)}$$

and the epimorphism $\theta_1 : \mathbb{G}_{R(\Omega^*)} \to \mathbb{G}_{R(\Omega_1^*)}$. In other words, the following diagram commutes:

$$\begin{array}{ccccc} \mathbb{G}_{R(\Omega^*)} & \longleftarrow & F[h_1, \ldots, h_{\rho_\Omega}] & \longrightarrow & F_{R(\Upsilon^*)} \\ \theta_1 \downarrow & & \tilde{\theta} \downarrow & & \downarrow \theta' \\ \mathbb{G}_{R(\Omega_1^*)} & \longleftarrow & F[h_1, \ldots, h_{\rho_{\Omega_1}}] & \longrightarrow & F_{R(\Upsilon_1^*)} \end{array}$$

PROOF. Follows by examining the definition of θ_1 for every derived transformation D1 − D6 and Lemma 4.6. □

LEMMA 4.12. *Using derived transformations, every generalised equation can be taken to the standard form.*

Let Ω_1 be obtained from Ω by an elementary or a derived transformation and let Ω be in the standard form. Then Ω_1 is in the standard form.

PROOF. Proof is straightforward. □

4.4. Construction of the tree $T(\Omega)$

In this section we describe a branching process for rewriting a generalised equation Ω. This process results in a locally finite and possibly infinite tree $T(\Omega)$. In the end of the section we describe infinite paths in $T(\Omega)$. We summarise the results of this section in the proposition below.

PROPOSITION 4.13. *For a (constrained) generalised equation $\Omega = \Omega_{v_0}$ over $\mathbb{F}$, one can effectively construct a locally finite, possibly infinite, oriented rooted at v_0 tree T, $T = T(\Omega_{v_0})$, such that:*

(1) *The vertices v_i of T are labelled by generalised equations Ω_{v_i} over $\mathbb{F}$.*
(2) *The edges $v_i \to v_{i+1}$ of T are labelled by epimorphisms*

$$\pi(v_i, v_{i+1}) : \mathbb{G}_{R(\Omega_{v_i}^*)} \to \mathbb{G}_{R(\Omega_{v_{i+1}}^*)}.$$

The edges $v_k \to v_{k+1}$, where v_{k+1} is a leaf of T and $\mathrm{tp}(v_{k+1}) = 1$, are labelled by proper epimorphisms. All the other epimorphisms $\pi(v_i, v_{i+1})$ are isomorphisms, in particular, edges that belong to infinite branches of the tree T are labelled by isomorphisms.

(3) *Given a solution H of Ω_{v_0}, there exists a path $v_0 \to v_1 \to \cdots \to v_l$ and a solution $H^{(l)}$ of Ω_{v_l} such that*

$$\pi_H = \pi(v_0, v_1) \cdots \pi(v_{l-1}, v_l) \pi_{H^{(l)}}.$$

Conversely, for every path $v_0 \to v_1 \to \cdots \to v_l$ in T and every solution $H^{(l)}$ of Ω_{v_l} the homomorphism

$$\pi(v_0, v_1) \cdots \pi(v_{l-1}, v_l) \pi_{H^{(l)}}$$

gives rise to a solution of Ω_{v_0}.

DEFINITION 4.14. Denote by $n_A = n_A(\Omega)$ the number of bases in the active sections of Ω and by ξ the number of open boundaries in the active sections.

For a closed section $\sigma \in \Sigma(\Omega)$ denote by $n(\sigma)$ the number of bases in σ. The *complexity* of a generalised equation Ω is defined as follows

$$\mathrm{comp} = \mathrm{comp}(\Omega) = \sum_{\sigma \in A\Sigma(\Omega)} \max\{0, n(\sigma) - 2\}.$$

REMARK 4.15. We use the following convention. Let Ω be a generalised equation. By a function $f(\Omega)$ of a generalised equation Ω we mean a function of the parameters $n_A(\Omega)$, $\xi(\Omega)$, ρ_Ω and $\mathrm{comp}(\Omega)$.

We begin with a general description of the tree $T(\Omega)$ and then construct it using induction on its height. For each vertex v in $T(\Omega)$ there exists a generalised equation Ω_v associated to v. Recall, that we consider only formally consistent generalised equations. The initial generalised equation Ω is associated to the root v_0, $\Omega_{v_0} = \Omega$. For each edge $v \to v'$ there exists a unique surjective homomorphism $\pi(v, v') : \mathbb{G}_{R(\Omega_v^*)} \to \mathbb{G}_{R(\Omega_{v'}^*)}$ associated to $v \to v'$.

If

$$v \to v_1 \to \ldots \to v_s \to u$$

is a path in $T(\Omega)$, then by $\pi(v, u)$ we denote the composition of corresponding homomorphisms

$$\pi(v, u) = \pi(v, v_1) \cdots \pi(v_s, u).$$

We call this epimorphism the *canonical homomorphism from $\mathbb{G}_{R(\Omega_v^*)}$ to $\mathbb{G}_{R(\Omega_u^*)}$*

There are two kinds of edges in $T(\Omega)$: *principal* and *auxiliary*. If not stated otherwise, every edge constructed is principal. Let $v \to v'$ be an edge of $T(\Omega)$, we assume that active (non-active) sections in $\Omega_{v'}$ are naturally inherited from Ω_v. If $v \to v'$ is a principal edge, then there exists a finite sequence of elementary and derived transformations from Ω_v to $\Omega_{v'}$ and the homomorphism $\pi(v, v')$ is a composition of the homomorphisms corresponding to these transformations.

Since both elementary and derived transformations uniquely associate a pair $(\Omega_i, H^{(i)})$ to (Ω, H):

$$\text{ET} : (\Omega, H) \to (\Omega_i, H^{(i)}) \text{ and } \text{D} : (\Omega, H) \to (\Omega_i, H^{(i)}),$$

a solution H of Ω defines a path in the tree $T(\Omega)$. We call such a path *the path defined by a solution H in T*.

Let Ω be a generalised equation. We construct an oriented rooted tree $T(\Omega)$. We start from the root v_0 and proceed by induction on the height of the tree.

Suppose, by induction, that the tree $T(\Omega)$ is constructed up to height n, and let v be a vertex of height n. We now describe how to extend the tree from v. The construction of the outgoing edges from v depends on which of the case described below takes place at the vertex v. We always assume that:

If the generalised equation Ω_v satisfies the assumptions of Case i, then Ω_v does not satisfy the assumptions of all the Cases j, with $j < i$.

The general guideline of the process is to use the entire transformation to transfer bases to the right of the interval. Before applying the entire transformation one has to make sure that the generalised equation is "clean". The stratification of the process into 15 cases, though seemingly unnecessary, is convenient in the proofs since each of the cases has a different behaviour with respect to "complexity" of the generalised equation, see Lemma 4.18.

Preprocessing. In Ω_{v_0} we transport closed sections using D 2 in such a way that all active sections are at the left end of the interval (the active part of the generalised equation), then come all non-active sections.

Termination conditions: Cases 1 and 2.

Case 1: The homomorphism $\pi(v_0, v)$ is not an isomorphism, or, equivalently, the canonical homomorphism $\pi(v_1, v)$, where $v_1 \to v$ is an edge of $T(\Omega)$, is not an isomorphism. The vertex v, in this case, is a leaf of the tree $T(\Omega)$. There are no outgoing edges from v.

Case 2: The generalised equation Ω_v does not contain active sections. Then the vertex v is a leaf of the tree $T(\Omega)$. There are no outgoing edges from v.

Moving constant bases to the non-active part: Cases 3 and 4.

Case 3: The generalised equation Ω_v contains a constant base λ in an active section such that the section $\sigma(\lambda)$ is not closed. In this case, we make the section $\sigma(\lambda)$ closed using the derived transformation D 1.

4.4. CONSTRUCTION OF THE TREE $T(\Omega)$

Case 4: The generalised equation Ω_v contains a constant base λ such that the section $\sigma(\lambda)$ is closed. In this case, we transport the section $\sigma(\lambda)$ to $C\Sigma$ using the derived transformation D 2. Suppose that the base λ is labelled by a letter $a \in \mathcal{A}^{\pm 1}$. Then we identify all closed sections of the type $[i, i+1]$, which contain a constant base with the label $a^{\pm 1}$, with the transported section $\sigma(\lambda)$, using the derived transformation D 6. In the resulting generalised equation $\Omega_{v'}$ the section $\sigma(\lambda)$ becomes a constant section, and the corresponding edge (v, v') is auxiliary, see Figure 8 (note that D 1 produces a finite set of generalised equations, Figure 8 shows only one of them).

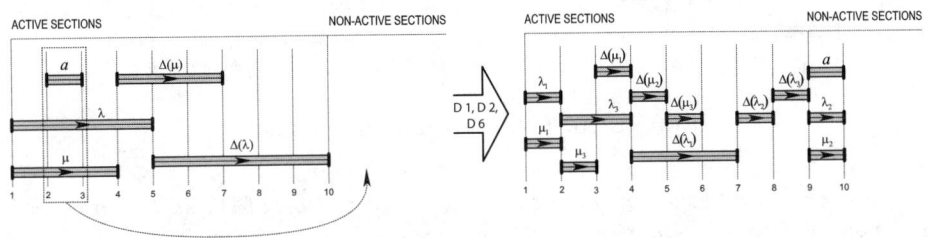

FIGURE 8. Cases 3-4: Moving constant bases.

Moving free variables to the non-active part: Cases 5 and 6.

Case 5: The generalised equation Ω_v contains a free variable h_q in an active section. Using D 2, we transport the section $[q, q+1]$ to the very end of the interval behind all the items of Ω_v. In the resulting generalised equation $\Omega_{v'}$ the transported section becomes a constant section, and the corresponding edge (v, v') is auxiliary.

Case 6: The generalised equation Ω_v contains a pair of matched bases in an active section. In this case, we perform ET 3 and delete it, see Figure 9.

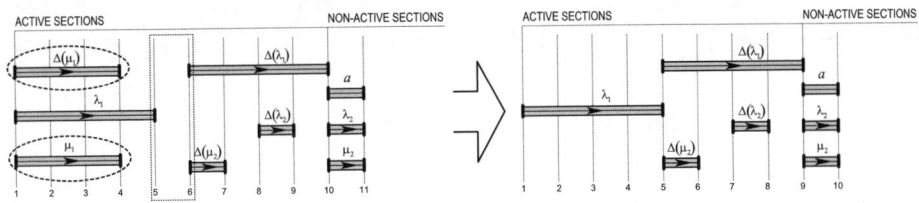

FIGURE 9. Cases 5-6: Removing a pair of matched bases and free variables.

REMARK 4.16. If Ω_v does not satisfy the conditions of any of the Cases 1-6, then the generalised equation Ω_v is in the standard form.

Eliminating linear variables: Cases 7-10.

Case 7: There exists an item h_i in an active section of Ω_v such that $\gamma_i = 1$ and such that both boundaries i and $i + 1$ are closed. Then, we remove the closed section $[i, i+1]$ together with the linear base using ET 4.

Case 8: There exists an item h_i in an active section of Ω_v such that $\gamma_i = 1$ and such that one of the boundaries i, $i+1$ is open, say $i+1$, and the other is closed. In this case, we first perform ET 5 and μ-tie $i+1$ by the only base μ it intersects; then using ET 1 we cut μ in $i+1$; and then we delete the closed section $[i, i+1]$ using ET 4, see Figure 10.

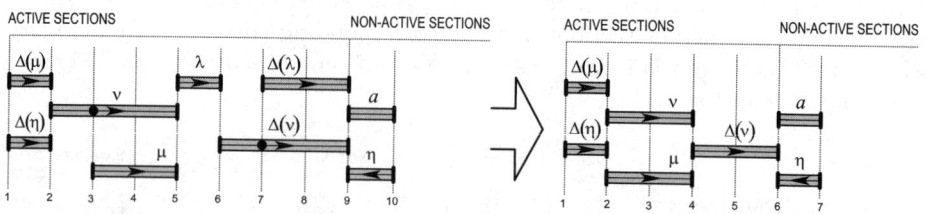

FIGURE 10. Cases 7-8: Linear variables.

Case 9: There exists an item h_i in an active section of Ω_v such that $\gamma_i = 1$ and such that both boundaries i and $i+1$ are open. In addition, there is a closed section σ such that σ contains exactly two bases μ_1 and μ_2, $\sigma = \sigma(\mu_1) = \sigma(\mu_2)$ and μ_1, μ_2 is not a pair of matched bases, i.e. $\mu_1 \neq \Delta(\mu_2)$; moreover, in the generalised equation $\widetilde{\Omega}_v = D\,3(\Omega)$ all the bases obtained from μ_1, μ_2 by ET 1 when constructing $\widetilde{\Omega}_v$ from Ω_v, do not belong to the kernel of $\widetilde{\Omega}_v$. In this case, using ET 5 we μ_1-tie all the boundaries that intersect μ_1; using ET 2, we transfer μ_2 onto $\Delta(\mu_1)$; and remove μ_1 together with the closed section σ using ET 4, see Figure 11.

Case 10: There exists an item h_i in an active section of Ω_v such that $\gamma_i = 1$ and such that both boundaries i and $i+1$ are open. In this event we close the section $[i, i+1]$ using D 1 and remove it using ET 4, see Figure 11.

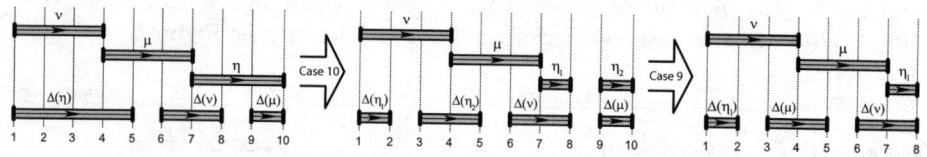

FIGURE 11. Cases 9-10: Linear case.

Tying a free boundary: Case 11.

Case 11: Some boundary i in the active part of Ω_v is free. Since the assumptions of Case 5 are not satisfied, the boundary i intersects at least one base, say, μ.

In this case, we μ-tie i using ET 5.

Quadratic case: Case 12.

Case 12: For every item h_i in the active part of Ω_v we have $\gamma_i = 2$. We apply the entire transformation D 5, see Figure 12.

Removing a closed section: Case 13.

4.4. CONSTRUCTION OF THE TREE $T(\Omega)$

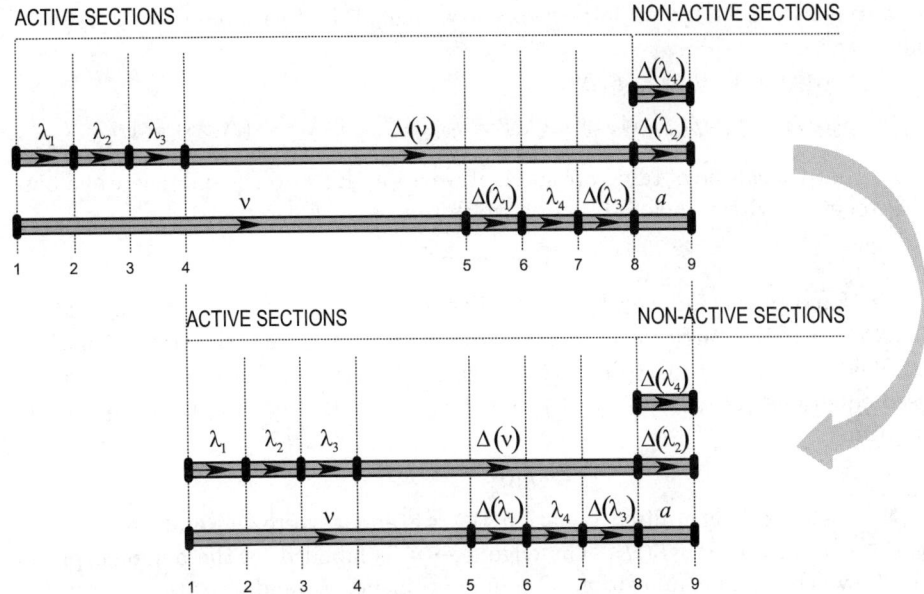

FIGURE 12. Case 12: Quadratic case, entire transformation.

Case 13: For every item h_i in the active part of Ω_v we have $\gamma_i \geq 2$, and $\gamma_{i'} > 2$ for at least one item $h_{i'}$ in the active part. Moreover, for some active base μ the section $\sigma(\mu)$ is closed. In this case, using D 2, we transport the section $\sigma(\mu)$ to the beginning of the interval (this makes μ a carrier base). We then apply the entire transformation D 5 (note that in this case the whole section $\sigma(\mu)$ is removed).

Tying a boundary: Case 14.

Case 14: For every item h_i in the active part of Ω_v we have $\gamma_i \geq 2$, and $\gamma_{i'} > 2$ for at least one item $h_{i'}$ in the active part. Moreover, some boundary j in the active part touches some base λ, intersects some base μ, and j is not μ-tied. In this case, using ET 5, we μ-tie j.

General case: Case 15.

Case 15: For every item h_i in the active part of Ω_v we have $\gamma_i \geq 2$, and $\gamma_{i'} > 2$ for at least one item $h_{i'}$ in the active part. Moreover, every active section $\sigma(\mu)$ is not closed and every boundary j is μ-tied in every base it intersects. We first apply the entire transformation D 5. Then, using ET 5, we μ-tie every boundary j in the active part that intersects a base μ and touches at least one base. This results in finitely many new vertices connected to v by principal edges.

If, in addition, Ω_v satisfies the assumptions of Case 15.1 below, then besides the principal edges already constructed, we construct a few more auxiliary edges outgoing from the vertex v.

Case 15.1: The carrier base μ of the generalised equation Ω_v intersects with its dual $\Delta(\mu)$. We first construct an auxiliary generalised equation $\widehat{\Omega}_v$ (which does *not* appear in the tree $T(\Omega)$) as follows. Firstly, we add a new constant section $[\rho_{\Omega_v}+1, \rho_{\Omega_v}+2]$ to the right of all the sections in Ω_v (in particular, $h_{\rho_{\Omega_v}+1}$ is a new

free variable). Secondly, we introduce a new pair of dual variable bases $\lambda, \Delta(\lambda)$ so that
$$\alpha(\lambda) = 1, \quad \beta(\lambda) = \beta(\Delta(\mu)),$$
$$\alpha(\Delta(\lambda)) = \rho_{\Omega_v} + 1, \quad \beta(\Delta(\lambda)) = \rho_{\Omega_v} + 2; \quad \varepsilon(\lambda) = \varepsilon(\Delta(\lambda)) = 1.$$

Notice that Ω_v can be obtained from $\widehat{\Omega}_v$ if we apply ET 4 to $\widehat{\Omega}_v$ and delete the base $\Delta(\lambda)$ together with the closed section $[\rho_v + 1, \rho_v + 2]$. Let

$$\hat{\pi}_v : \mathbb{G}_{R(\Omega_v^*)} \to \mathbb{G}_{R(\widehat{\Omega}_v^*)}$$

be the isomorphism induced by ET 4. The assumptions of Case 15 still hold for $\widehat{\Omega}_v$. Note that the carrier base of $\widehat{\Omega}_v$ is the base λ. Applying the transformations described in Case 15 to $\widehat{\Omega}_v$ (first D 5 and then ET 5), we obtain a set of new generalised equations $\{\Omega_{v'_i} \mid i = 1, \ldots, \mathfrak{n}\}$ and the set of corresponding epimorphisms of the form:

$$\phi_{v'_i} : \mathbb{G}_{R(\widehat{\Omega}_v^*)} \to \mathbb{G}_{R(\Omega_{v'_i}^*)}.$$

Now for each generalised equation $\Omega_{v'_i}$ we add a vertex v'_i and an auxiliary edge $v \to v'_i$ in the tree $T(\Omega)$. The edge $v \to v'_i$ is labelled by the homomorphism $\pi(v, v'_i)$ which is the composition of homomorphisms $\phi_{v'_i}$ and $\hat{\pi}_v$, $\pi(v, v'_i) = \hat{\pi}_v \phi_{v'_i}$. We associate the generalised equation $\Omega_{v'_i}$ to the vertex v'_i.

The tree $T(\Omega)$ is therefore constructed. Observe that, in general, $T(\Omega)$ is an infinite locally finite tree.

If the generalised equation Ω_v satisfies the assumptions of Case i ($1 \le i \le 15$), then we say that the vertex v has *type* i and write $\text{tp}(v) = i$.

LEMMA 4.17. *For any vertex v of the tree $T(\Omega)$ there exists an algorithm to decide whether or not v is of type k, $k = 1, \ldots, 15$.*

PROOF. The statement of the lemma is obvious for all cases, except for Case 1.

To decide whether or not the vertex v is of type 1 it suffices to show that there exists an algorithm to check whether or not the canonical epimorphism $\pi(v, u)$ associated to an edge $v \to u$ in $T(\Omega)$ is a proper epimorphism. This can be done effectively by Lemma 4.5. □

LEMMA 4.18 (cf. Lemma 3.1, [**Raz87**]). *Let $u \to v$ be a principal edge of the tree $T(\Omega)$. Then the following statements hold.*

(1) *If $\text{tp}(u) \ne 3, 10$, then $n_A(\Omega_v) \le n_A(\Omega_u)$, moreover if $\text{tp}(u) = 6, 7, 9, 13$, then this inequality is strict;*
(2) *If $\text{tp}(u) = 10$, then $n_A(\Omega_v) \le n_A(\Omega_u) + 2$;*
(3) *If $\text{tp}(u) \le 13$ and $\text{tp}(u) \ne 3, 11$, then $\xi(\Omega_v) \le \xi(\Omega_u)$;*
(4) *If $\text{tp}(u) \ne 3$, then $\text{comp}(\Omega_v) \le \text{comp}(\Omega_u)$.*

PROOF. Straightforward verification. □

The following lemma gives a description of the infinite branches in the tree $T(\Omega)$.

LEMMA 4.19 (cf. Lemma 3.2, [**Raz87**]). *Let*

(4.6) $$v_0 \to v_1 \to \ldots \to v_r \to \ldots$$

be an infinite path in the tree $T(\Omega)$. Then there exists a natural number N such that all the edges $v_n \to v_{n+1}$ of this path with $n \geq N$ are principal edges, and one of the following conditions holds:

(A)   linear case: $7 \leq \mathrm{tp}(v_n) \leq 10$ for all $n \geq N$;
(B)   quadratic case: $\mathrm{tp}(v_n) = 12$ for all $n \geq N$;
(C)   general case: $\mathrm{tp}(v_n) = 15$ for all $n \geq N$.

PROOF. Firstly, note that Cases 1 and 2 can only occur once in the path (4.6).

Secondly, note Cases 3 and 4 can occur only finitely many times in the path (4.6), namely, at most $2t$ times where t is the number of constant bases in the original generalised equation Ω. Therefore, there exists a natural number N_1 such that $\mathrm{tp}(v_i) \geq 5$ for all $i \geq N_1$.

Now we show that the number of vertices v_i ($i \geq N$) for which $\mathrm{tp}(v_i) = 5$ is bounded above by the minimal number of generators of the group $\mathbb{G}_{R(\Upsilon^*)}$, in particular, it cannot be greater than $\rho + 1 + |\mathcal{A}|$, where $\rho = \rho_\Upsilon$. Indeed, if a path from the root v_0 to a vertex v contains k vertices of type 5, then Υ_v has at least k free variables in the non-active part. This implies that the coordinate group $\mathbb{G}_{R(\Upsilon_v^*)}$ has the free group of rank k as a free factor, hence it cannot be generated by less than k elements. Since $\pi'(v_0,v) : \mathbb{G}_{R(\Upsilon^*)} \to \mathbb{G}_{R(\Upsilon_v^*)}$ is a surjective homomorphism, the group $\mathbb{G}_{R(\Upsilon^*)}$ cannot be generated by less then k elements. This shows that $k \leq \rho + 1 + |\mathcal{A}|$. It follows that there exists a number $N_2 \geq N_1$ such that $\mathrm{tp}(v_i) > 5$ for every $i \geq N_2$.

Since the path (4.6) is infinite, by Lemma 4.18, we may assume that for every $i,j \geq N_3$ one has $\mathrm{comp}(\Omega_{v_i}) = \mathrm{comp}(\Omega_{v_j})$, i.e. the complexity stabilises.

If $v_i \to v_{i+1}$ is an auxiliary edge, where $i \geq N_3$, then $\mathrm{tp}(v_i) = 15$ and Ω_{v_i} satisfies the assumption of Case 15.1. In the notation of Case 15.1, we analyse the complexity of the generalised equation $\Omega_{v_{i+1}}$ obtained from $\widehat{\Omega}_{v_i}$. As both bases μ and $\Delta(\mu)$ are transferred from the carrier base λ of $\widehat{\Omega}_{v_i}$ to the non-active part, so the complexity decreases by at least two, i.e. $\mathrm{comp}(\Omega_{v_{i+1}}) \leq \mathrm{comp}(\widehat{\Omega}_{v_i}) - 2$. Observe also that $\mathrm{comp}(\widehat{\Omega}_{v_i}) = \mathrm{comp}(\Omega_{v_i}) + 1$. Hence $\mathrm{comp}(\Omega_{v_{i+1}}) < \mathrm{comp}(\Omega_{v_i})$, which derives a contradiction. Hence, if $i \geq N_3$ the edge $v_i \to v_{i+1}$ is principal.

Suppose that $i \geq N_3$. If $\mathrm{tp}(v_i) = 6$, then the closed section containing the matched bases $\mu, \Delta(\mu)$, does not contain any other bases (otherwise $\mathrm{comp}(\Omega_{v_{i+1}}) < \mathrm{comp}(\Omega_{v_i})$). But then $\mathrm{tp}(v_{i+1}) = 5$ deriving a contradiction. We therefore get that for all $i \geq N_3$ one has $\mathrm{tp}(v_i) > 6$.

If $\mathrm{tp}(v_i) = 12$, then it is easy to see that $\mathrm{tp}(v_{i+1}) = 6$ or $\mathrm{tp}(v_{i+1}) = 12$. Therefore, if $i \geq N_3$ we get that $\mathrm{tp}(v_{i+1}) = \mathrm{tp}(v_{i+2}) = \cdots = \mathrm{tp}(v_{i+j}) = 12$ for every $j > 0$ and we have case (B) of the lemma.

Suppose that $\mathrm{tp}(v_i) \neq 12$ for all $i \geq N_3$.

Notice that the only elementary (derived) transformation that may produce free boundaries is ET 3 and that ET 3 is applied only in Case 6. Since for $i \geq N_3$ we have $\mathrm{tp}(v_i) \geq 7$, we see that there are no new free boundaries in the generalised equations Ω_{v_i} for $i \geq N_3$. It follows that there exists a number $N_4 \geq N_3$ such that $\mathrm{tp}(v_i) \neq 11$ for every $i \geq N_4$.

Suppose now that for some $i \geq N_4$, $13 \leq \mathrm{tp}(v_i) \leq 15$. It is easy to see from the description of Cases 13, 14 and 15 that $\mathrm{tp}(v_{i+1}) \in \{6, 13, 14, 15\}$. Since $\mathrm{tp}(v_{i+1}) \neq 6$, this implies that $13 \leq \mathrm{tp}(v_j) \leq 15$ for every $j \geq i$. Then, by Lemma 4.18, we have that $n_A(\Omega_{v_j}) \leq n_A(\Omega_{v_{N_4}})$, for every $j \geq N_4$. Furthermore, if $\text{tp}(v_j) = 13$, then $n_A(\Omega_{v_{j+1}}) < n_A(\Omega_{v_j})$. Hence there exists a number $N_5 \geq N_4$ such that $\text{tp}(v_j) \neq 13$ for all $j \geq N_5$.

Suppose that $i \geq N_5$. Note that there can be at most $8(n_A(\Omega_{v_i}))^2$ vertices of type 14 in a row starting at the vertex v_i, hence there exists $j \geq i$ such that $\text{tp}(v_j) = 15$. From the description of Case 15 it follows that $\text{tp}(v_{j+1}) \neq 14$, thus $\text{tp}(v_{j+1}) = 15$, and we have case (C) of the lemma.

Finally, we are left with the case when $7 \leq \text{tp}(v_i) \leq 10$ for all the vertices of the path. We then have case (A) of the lemma. □

REMARK 4.20. Let
$$(\Omega_{v_1}, H^{(1)}) \to (\Omega_{v_2}, H^{(2)}) \to \cdots \to (\Omega_{v_l}, H^{(l)})$$
be the path defined by the solution $H^{(1)}$. If $\text{tp}(v_i) \in \{7, 8, 9, 10, 12, 15\}$, then by Lemma 4.10, $|H^{(i+1)}| < |H^{(i)}|$.

CHAPTER 5

Minimal solutions

In this chapter we introduce a reflexive, transitive relation on the set of solutions of a generalised equation. We use this relation to introduce the notion of a minimal solution with respect to a group of automorphisms of the coordinate group of the generalised equation. In the end of the section we describe the behaviour of minimal solutions with respect to the elementary transformations.

Let u and v be two geodesic words from $\mathbb{G}$. Consider the product uv of u and v. The geodesic $\overline{uv}$ of the element uv can be written as follows $\overline{uv} \doteq u_1 \cdot v_2$, where $u = u_1 d^{-1}$, $v = d v_2$ and d is the greatest (left) common divisor of u^{-1} and v (in the sense of [**EKR05**]). Notice that if the word uv is not geodesic, then it follows that d is non-trivial. We call d the *cancellation divisor of u and v* and we denote it by $\mathrm{CD}(u,v)$.

DEFINITION 5.1. Let $\mathbb{G}(\mathcal{A} \cup B)$ be a partially commutative $\mathbb{G}(\mathcal{A})$-group (i.e. the subgroup of $\mathbb{G}(\mathcal{A} \cup B)$ generated by $\mathcal{A}$ is isomorphic to $\mathbb{G}(\mathcal{A})$). Let Ω be a generalised equation with coefficients from $(\mathcal{A} \cup B)^{\pm 1}$ over the free monoid $\mathbb{F}(\mathcal{A}^{\pm 1} \cup B^{\pm 1})$. Let $\mathfrak{C}(\Omega)$ be an arbitrary group of $\mathbb{G}(\mathcal{A})$-automorphisms of $\mathbb{G}(\mathcal{A} \cup B)[h]/R(\Omega^*)$. For solutions $H^{(1)}$ and $H^{(2)}$ of the generalised equation Ω we write $H^{(1)} <_{\mathfrak{C}(\Omega)} H^{(2)}$ if there exists a $\mathbb{G}(\mathcal{A})$-endomorphism π of the group $\mathbb{G}(\mathcal{A} \cup B)$ and an automorphism $\sigma \in \mathfrak{C}(\Omega)$ such that the following conditions hold:

(1) $\pi_{H^{(2)}} = \sigma \pi_{H^{(1)}} \pi$;
(2) for any k, j, $1 \leq k, j \leq \rho$, if the word $(H_k^{(2)})^\epsilon (H_j^{(2)})^\delta$, $\epsilon, \delta \in \{1, -1\}$, is geodesic as written, then the word $(H_k^{(1)})^\epsilon (H_j^{(1)})^\delta$ is geodesic as written;
(3) for any k, j, $1 \leq k, j \leq \rho$ such that the word $(H_k^{(2)})^\epsilon (H_j^{(2)})^\delta$, $\epsilon, \delta \in \{1, -1\}$, is not geodesic, if

(5.1) $\quad \mathbb{A}(J) \cap \left\{ a \in \mathcal{A}^{\pm 1} \mid a \text{ is a left divisor of } \mathrm{CD}\left((H_k^{(2)})^\epsilon, (H_j^{(2)})^\delta \right) \right\} = \emptyset,$

for some $J = \{H_{i_1}^{(2)}, \ldots, H_{i_t}^{(2)}\}$, then either the word $(H_k^{(1)})^\epsilon (H_j^{(1)})^\delta$ is geodesic as written or

(5.2) $\quad \mathbb{A}(J') \cap \left\{ a \in \mathcal{A}^{\pm 1} \mid a \text{ is a left divisor of } \mathrm{CD}\left((H_k^{(1)})^\epsilon, (H_j^{(1)})^\delta \right) \right\} = \emptyset,$

where $J' = \{H_{i_1}^{(1)}, \ldots, H_{i_t}^{(1)}\}$.

Obviously, the relation '$<_{\mathfrak{C}(\Omega)}$' is transitive. We would like to draw the reader's attention to the fact that the relation $H <_{\mathfrak{C}(\Omega)} H'$ does not imply relations on the lengths of the solutions H and H'.

The motivation for property (3) in the above definition is the following lemma.

LEMMA 5.2. *Let $\mathbb{G}(\mathcal{A} \cup \mathcal{B})$ be a partially commutative $\mathbb{G}$-group, let Ω be a generalised equation with coefficients from $(\mathcal{A} \cup \mathcal{B})^{\pm 1}$ over the free monoid $\mathbb{F}(\mathcal{A}^{\pm 1} \cup \mathcal{B}^{\pm 1})$. Let $\mathfrak{C}(\Omega)$ be an arbitrary group of $\mathbb{G}(\mathcal{A})$-automorphisms of $\mathbb{G}(\mathcal{A} \cup \mathcal{B})/R(\Omega^*)$ and let $H^{(1)}$ and $H^{(2)}$ be two solutions of the generalised equation Ω such that $H^{(1)} <_{\mathfrak{C}(\Omega)} H^{(2)}$. Then for any word $W(x_1, \ldots, x_\rho) \in F(x_1, \ldots, x_\rho)$ such that the $W(H_1^{(2)}, \ldots, H_\rho^{(2)})$ is geodesic as written (treated as an element of $\mathbb{G}(\mathcal{A} \cup \mathcal{B})$), the word $W(H_1^{(1)}, \ldots, H_\rho^{(1)})$ is geodesic as written (treated as an element of $\mathbb{G}(\mathcal{A} \cup \mathcal{B})$).*

PROOF. We use induction on the length of W. If $|W| = 1$, then the statement follows, since by the definition of a solution of the generalised equation, the words $(H_j^{(1)})^{\pm 1}$ and $(H_j^{(2)})^{\pm 1}$ are both geodesic.

Let $W(x_1, \ldots, x_\rho) = x_{l_1}^{\epsilon_1} \cdots x_{l_{n+1}}^{\epsilon_{n+1}}$, $\epsilon_i \in \{-1, 1\}$, $1 \leq i \leq n+1$, be a word such that $W(H_1^{(2)}, \ldots, H_\rho^{(2)})$ is geodesic and assume that $W(H_1^{(1)}, \ldots, H_\rho^{(1)})$ is not geodesic. By induction hypothesis, the word $W_n(H^{(1)}) = H_{l_1}^{(1)\epsilon_1} \cdots H_{l_n}^{(1)\epsilon_n}$ is geodesic. Consider the cancellation divisor $D = \mathrm{CD}\left(W_n(H^{(1)}), H_{l_{n+1}}^{(1)\ \epsilon_{n+1}}\right)$. Analogously, by induction hypothesis, the word $H_{l_2}^{(1)\epsilon_2} \cdots H_{l_{n+1}}^{(1)\ \epsilon_{n+1}}$ is geodesic. Using the fact that the word $W_n(H^{(1)})$ is geodesic, we get that the cancellation divisor $\mathrm{CD}\left(H_{l_1}^{(1)\epsilon_1}, H_{l_2}^{(1)\epsilon_2} \cdots H_{l_{n+1}}^{(1)\ \epsilon_{n+1}}\right)$ is, in fact, D, and, in particular, no left divisor of the word $H_{l_2}^{(1)\epsilon_2} \cdots H_{l_n}^{(1)\epsilon_n}$ left-divides D. We thereby have that

(a) on the one hand, the word $H_{l_1}^{(1)\epsilon_1} D$ is not geodesic and,

(b) on the other hand, by Proposition 3.18 of [**EKR05**], that $D \leftrightarrows H_{l_2}^{(1)\epsilon_2} \cdots H_{l_n}^{(1)\epsilon_n}$.

From (a) we get that the word $H_{l_1}^{(1)\epsilon_1} H_{l_{n+1}}^{(1)\ \epsilon_{n+1}}$ is not geodesic, and thus by property (2) from Definition 5.6, it follows that the word $H_{l_1}^{(2)\epsilon_1} H_{l_{n+1}}^{(2)\ \epsilon_{n+1}}$ is not geodesic.

From (b) and the fact that the word $H_{l_2}^{(1)\epsilon_2} \cdots H_{l_n}^{(1)\epsilon_n}$ is geodesic, we have that $D \in \mathbb{A}(H_{l_i}^{(1)})$ for every $i = 2, \ldots, n$. It follows that Equation (5.2) fails for the set $J' = \{H_{l_2}^{(1)}, \ldots, H_{l_n}^{(1)}\}$ and thus Equation (5.1) fails for the set $J = \{H_{l_2}^{(2)}, \ldots, H_{l_n}^{(2)}\}$. This derives a contradiction with the fact that $H_{l_1}^{(2)\epsilon_1} \cdots H_{l_{n+1}}^{(2)\ \epsilon_{n+1}}$ is geodesic, since there exists a letter $a \in \mathcal{A}^{\pm 1}$ such that a is a divisor of $\mathrm{CD}\left(H_{l_1}^{(2)\epsilon_1}, H_{l_{n+1}}^{(2)\ \epsilon_{n+1}}\right)$ and $a \in \mathbb{A}(J)$. □

LEMMA 5.3. *Let the generalised equation Ω_1 be obtained from the generalised equation Ω by one of the elementary transformation $\mathrm{ET}\, 1 - \mathrm{ET}\, 5$, i.e. $\Omega_1 \in \mathrm{ET}\, i(\Omega)$ for some $i = 1, \ldots, 5$. Let H be a solution of Ω and $H^{(1)}$ be a solution of Ω_1 so that the following diagram commutes*

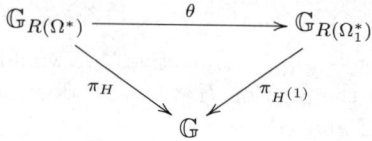

5. MINIMAL SOLUTIONS

Let $H^{(1)+}$ be another solution of Ω_1 so that $H^{(1)+} <_{\mathfrak{C}(\Omega)} H^{(1)}$, where $\mathfrak{C}(\Omega)$ is arbitrary. The ρ-tuple H^+ of geodesic words of $\mathbb{G}$ defined by the homomorphism $\pi_{H^+} = \theta \pi_{H^{(1)+}}$ is a solution of the generalised equation Ω.

PROOF. Proof is by examination of the definitions of elementary transformations. We further use the notation introduced in the definitions of elementary transformations.

Suppose first that Ω_1 is obtained from Ω by ET 1. Since, in this case, $H^+ = H^{(1)+}$, we have that H^+ is a tuple of non-trivial geodesic words in $\mathbb{G}$. The elementary transformation ET 1 is invariant on all the bases but the pair λ, $\Delta(\lambda)$. Therefore, we are left to prove that the words $H^+(\lambda)$ and $H^+(\Delta(\lambda))$ are geodesic and that the equality $H^+(\lambda)^{\varepsilon(\lambda)} = H^+(\Delta(\lambda))^{\varepsilon(\Delta(\lambda))}$ is graphical.

Since $H(\lambda) = H^{(1)}[\alpha(\lambda_1), \beta(\lambda_2)]$ is a geodesic word and $H^{(1)+} <_{\mathfrak{C}(\Omega)} H^{(1)}$, by Lemma 5.2 it follows that $H^+(\lambda) = H^{(1)+}[\alpha(\lambda_1), \beta(\lambda_2)]$ is geodesic. Similarly, $H^+(\Delta(\lambda))$ is also geodesic. Since $H^{(1)+}$ is a solution of Ω_1, we have that

$$H^{(1)+}(\lambda_i)^{\varepsilon(\lambda_i)} \doteq H^{(1)+}(\Delta(\lambda_i))^{\varepsilon(\Delta(\lambda_i))}, \quad \text{where } i = 1, 2.$$

From the equalities

$$H^+(\lambda) = H^{(1)+}(\lambda_1) H^{(1)+}(\lambda_2) \quad \text{and} \quad H^+(\Delta(\lambda)) = H^{(1)+}(\Delta(\lambda_1)) H^{(1)+}(\Delta(\lambda_2)),$$

and the fact that the words $H^+(\lambda)$ and $H^+(\Delta(\lambda))$ are geodesic, we obtain that the equality $H^+(\lambda)^{\varepsilon(\lambda)} = H^+(\Delta(\lambda))^{\varepsilon(\Delta(\lambda))}$ is graphical.

The other cases are similar and left to the reader. □

DEFINITION 5.4. Let Ω be a generalised equation in ρ variables and let H be a solution of Ω. Consider a $2\rho \times 2\rho$ matrix (m_{i_1,i_2}), $1 \le i_1, i_2 \le 2\rho$ constructed by the solution H in the following way. The elements m_{i_1,i_2} of the matrix are $2^\rho + 1$-vectors with entries from the set $\{0, 1\}$. We enumerate (in an arbitrary way) the set $\chi(\{H_1, \ldots, H_\rho\})$ of all subsets of $\{H_1, \ldots, H_\rho\}$. Abusing the notation, in this definition if $i_l > \rho$, then by H_{i_l} we mean $H_{i_l - \rho}^{-1}$.

The vector m_{i_1,i_2} has all of its components equal to 0 if and only if the word $H_{i_1} H_{i_2}$ is geodesic as written. The first component of the vector m_{i_1,i_2} equals 1 if and only if the word $H_{i_1} H_{i_2}$ is not geodesic. The l-th component of m_{i_1,i_2} equals 1, $l > 1$, if for the $l-1$-th set J of $\chi(\{H_1, \ldots, H_\rho\})$ Equation (5.1) is satisfied. Otherwise the l-th component of m_{i_1,i_2} equals 0.

We call the matrix (m_{i_1,i_2}), $1 \le i_1, i_2 \le 2\rho$ the *cancellation matrix* of the solution H.

REMARK 5.5. Note that every generalised equation Ω with coefficients from $\mathbb{G}(\mathcal{A})$ can be considered as a generalised equation Ω_B with coefficients from $\mathbb{G}(\mathcal{A} \cup B)$ for *any* partially commutative $\mathbb{G}(\mathcal{A})$-group $\mathbb{G}(\mathcal{A} \cup B)$. Furthermore, any solution H of Ω induces a solution H^B of Ω_B so that the following diagram commutes:

$$\begin{array}{ccc} \mathbb{G}_{R(\Omega^*)} & \hookrightarrow & \mathbb{G}(\mathcal{A} \cup B)[h]/_{R(\Omega_B^*)} \\ \pi_H \downarrow & & \downarrow \pi_{H^B} \\ \mathbb{G} & \hookrightarrow & \mathbb{G}(\mathcal{A} \cup B) \end{array}$$

Similarly, any group of $\mathbb{G}(\mathcal{A})$-automorphisms $\mathfrak{C}(\Omega)$ of $\mathbb{G}_{R(\Omega^*)}$ defines a group of $\mathbb{G}(\mathcal{A} \cup \mathcal{B})$-automorphisms $\mathfrak{C}(\Omega_B)$ of $\mathbb{G}(\mathcal{A} \cup \mathcal{B})[h]/R(\Omega_B^*)$.

DEFINITION 5.6. In the above notation, a solution H of Ω is called *minimal with respect to the group of automorphisms* $\mathfrak{C}(\Omega)$ if there do not exist a partially commutative group $\mathbb{G}(\mathcal{A} \cup \mathcal{B})$ and a solution H' of the generalised equation Ω_B so that $H' <_{\mathfrak{C}(\Omega_B)} H^B$ and $|H'_k| \le |H_k^B|$ for all k, $k = 1, \ldots, \rho$ and $|H_j| < |H_j^B|$ for at least one j, $1 \le j \le \rho$.

Since the length of a solution H is a positive integer, every strictly descending chain of solutions
$$H >_{\mathfrak{C}(\Omega)} H^{(1)} >_{\mathfrak{C}(\Omega)} \cdots >_{\mathfrak{C}(\Omega)} H^{(k)} >_{\mathfrak{C}(\Omega)} \cdots$$
is finite. It follows that for every solution H of Ω there exists a minimal solution H^+ such that $H^+ <_{\mathfrak{C}(\Omega)} H$.

REMARK 5.7. Note that given a solution H, a minimal solution H' so that $H' <_{\mathfrak{C}(\Omega)} H$ is not unique, i.e. there may exist minimal solutions H' and H^+ so that $H' <_{\mathfrak{C}(\Omega)} H^+$ and $H^+ <_{\mathfrak{C}(\Omega)} H'$, but, on the one hand, one has $|H'_k| < |H_k^+|$ for some $1 \le k \le \rho$ and, on the other hand, $|H'_j| > |H_j^+|$ for some $1 \le j \le \rho$.

The reason for extending the generating set from $\mathcal{A}$ to $\mathcal{A} \cup \mathcal{B}$ in the definition above, becomes clear in the proof of Lemma 6.19

LEMMA 5.8. *Let the generalised equation Ω_1 be obtained from the generalised equation Ω by one of the elementary transformation* ET 1 $-$ ET 5, *i.e.* $\Omega_1 \in \mathrm{ET}\,i(\Omega)$ *for some* $i = 1, \ldots, 5$ *and suppose that the corresponding epimorphism θ from $\mathbb{G}_{R(\Omega^*)}$ to $\mathbb{G}_{R(\Omega_1^*)}$ is an isomorphism. Let H be a solution of Ω and $H^{(1)}$ be a solution of Ω_1 so that the following diagram commutes*

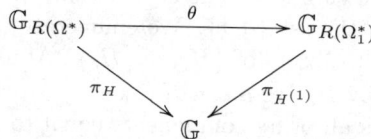

If H is a minimal solution of Ω with respect to a group of automorphisms $\mathfrak{C}$ of $\mathbb{G}_{R(\Omega^*)}$, then $H^{(1)}$ is a minimal solution of Ω_1 with respect to the group of automorphisms $\theta^{-1}\mathfrak{C}\theta$.

PROOF. Assume the converse, i.e. $H^{(1)}$ is not minimal with respect to $\theta^{-1}\mathfrak{C}\theta$. Then there exists a solution $H^{(1)+}$ of Ω_1 so that $H^{(1)+} <_{\theta^{-1}\mathfrak{C}\theta} H^{(1)}$, $|H^{(1)+}_k| \le |H^{(1)}_k|$ for all k and $|H^{(1)+}_j| < |H^{(1)}_j|$ for some j.

Let H^+ be a solution of the system of equations Ω^* so that $\pi_{H^+} = \theta \pi_{H^{(1)+}}$. By Lemma 5.3, H^+ is a solution of Ω.

By Lemma 5.2, condition (2) from Definition 5.1 holds for the pair H^+ and H. We now show that $H^+ <_{\mathfrak{C}} H$ and derive a contradiction with the minimality of H.

We now show that condition (3) from Definition 5.1 holds. Assume that for the pair H_k^+, H_j^+ the word $H_k^{+\epsilon} H_j^{+\delta}$ is not geodesic and that there exists a set $J' = \{H_{i_1}^+, \ldots, H_{i_t}^+\}$ such that Equation (5.2) fails, i.e. there exists a set J' and a letter $a \in \mathcal{A}^{\pm 1}$ such that $a \in \mathbb{A}(J')$ and a is a left divisor of $\mathrm{CD}(H_k^{+\epsilon}, H_j^{+\delta})$.

5. MINIMAL SOLUTIONS

Without loss of generality, we further assume that $\epsilon, \delta = 1$. Write H_k^+ and H_j^+ as words in the $H_i^{(1)+}$'s:

$$H_k^+ \doteq H_{k_1}^{(1)+\epsilon_1} \cdots H_{k_s}^{(1)+\epsilon_s} \text{ and } H_j^+ \doteq H_{j_1}^{(1)+\delta_1} \cdots H_{j_r}^{(1)+\delta_r}.$$

It follows that a is a left divisor of $\mathrm{CD}\left(H_{k_m}^{(1)+\epsilon_m}, H_{j_n}^{(1)+\delta_n}\right)$ for some m and n, and that

$$a \leftrightharpoons H_{j_1}^{(1)+\delta_1} \cdots H_{j_{n-1}}^{(1)+\delta_{n-1}} \text{ and } a \leftrightharpoons H_{k_{m+1}}^{(1)+\epsilon_{m+1}} \cdots H_{k_s}^{(1)+\epsilon_s}.$$

Therefore, Equation (5.2) fails for the solution $H^{(1)+}$ and the set $J^{(1)'}$, where the set $J^{(1)'}$ is a union of words $H_i^{(1)+}$ that appear in the decomposition of a word $H_{i_l}^+ \in J'$ and the set $\{H_{j_1}^{(1)+}, \ldots, H_{j_{n-1}}^{(1)+}, H_{k_{m+1}}^{(1)+}, \ldots, H_{k_s}^{(1)+}\}$. Since $H^{(1)+} <_{\theta^{-1}\mathfrak{C}\theta} H^{(1)}$, by condition (3) from Definition 5.1, Equation (5.1) fails for the solution $H^{(1)}$ and the corresponding set $J^{(1)}$, i.e. there exists a letter $b \in \mathcal{A}^{\pm 1}$ such that $b \in \mathbb{A}(J^{(1)})$ and b is a left divisor of $\mathrm{CD}\left(H_{k_m}^{(1)\epsilon_m}, H_{j_n}^{(1)\delta_n}\right)$.

Since $b \leftrightharpoons \{H_{j_1}^{(1)}, \ldots, H_{j_{n-1}}^{(1)}, H_{k_{m+1}}^{(1)}, \ldots, H_{k_s}^{(1)}\}$, we get that b is a left divisor of $\mathrm{CD}(H_k^\epsilon, H_j^\delta)$. Furthermore, since $b \leftrightharpoons$-commutes with the words $H_i^{(1)}$'s that appear in the decomposition of a word $H_{i_l} \in J$ (where J is the set corresponding to J'), we get that $b \in \mathbb{A}(J)$. It follows that Equation (5.1) fails for the solution H and the set J. Hence, condition (3) from Definition 5.1 holds for the solutions H and H^+.

Furthermore, by condition (1) from Definition 5.1, we have $\pi_{H^{(1)}} = \theta^{-1}\psi\theta\pi_{H^{(1)+}}$, where $\psi \in \mathfrak{C}$, hence $\pi_H = \psi\pi_{H^+}$, and thus condition (1) from Definition 5.1 holds for the pair H^+ and H. We thereby have proven that $H^+ <_{\mathfrak{C}} H$.

Finally, since $H_i \doteq w_i(H^{(1)})$, $H_i^+ \doteq w_i(H^{(1)+})$ and $|H^{(1)+}_k| \leq |H_k^{(1)}|$ for all k and $|H^{(1)+}_j| < |H_j^{(1)}|$ for some j, we get that $|H_k^+| \leq |H_k|$ for all k and $|H_l^+| < |H_l|$ for some l, contradicting the minimality of H. $\square$

CHAPTER 6

Periodic structures

Informally, the aim of this chapter is to prove the following strong version of the so-called Bulitko's Lemma, [**Bul70**]:

Applying automorphisms from a finitely generated subgroup $\mathfrak{A}(\Omega)$ of the group of automorphisms of the coordinate group $\mathbb{G}_{R(\Omega^*)}$ to a periodic solution either one can bound the exponent of periodicity of the solution (regular case, see Lemma 6.20), *or one can get a solution of a proper equation (strongly singular case, see* Lemma 6.17, *and singular case, see* Lemma 6.18).

Above, by a solution of a proper equation we mean a homomorphism from a proper quotient of the coordinate group of Ω to $\mathbb{G}$.

This approach for free groups was introduced by A. Razborov. In [**Raz85**], he defines a combinatorial object, called a periodic structure on a generalised equation Ω and constructs a finite set of generators for the group $\mathfrak{A}(\Omega)$.

In Section 6.2 we give an example that follows the exposition of Section 6.1. We advise the reader unfamiliar with the definitions, to consult this example while reading Section 6.1.

6.1. Periodic structures

We fix till the end of this section a generalised equation $\Omega = \langle \Upsilon, \mathfrak{R}_\Upsilon \rangle$ in the standard form. Suppose that some boundary k (between h_{k-1} and h_k) in the active part of Ω does not touch bases. Since the generalised equation Ω is in the standard form, the boundary k intersects at least one base μ. Using ET 5 we μ-tie the boundary k. Applying D 3, if necessary, we may assume that the set of boundary connections in Ω is empty and that each boundary of Ω touches a base.

A cyclically reduced word P in $\mathbb{G}$ is called a *period* if P is geodesic (in $\mathbb{G}$) and is not a proper power treated as an element of the ambient free monoid. A word w is called *P-periodic* if w is geodesic (treated as an element of $\mathbb{G}$), $|w| \geq |P|$ and, w is a subword of P^n for some n. Every P-periodic word w can be presented in the form

(6.1) $$w \doteq Q^r Q_1$$

where Q is a cyclic permutation of $P^{\pm 1}$, $r \geq 1$, $Q \doteq Q_1 Q_2$ and $Q_2 \neq 1$. The number r is called the *exponent* of w. A maximal exponent of a P-periodic subword in a word w is called the *exponent of P-periodicity of w*. We denote it by $\exp(w)$.

DEFINITION 6.1. Let Ω be a generalised equation in the standard form. A solution $H = (H_1, \ldots, H_\rho)$ of Ω is called *periodic with respect to a period P*, if for every closed variable section σ of Ω one of the following conditions holds:

(1) $H(\sigma)$ is P-periodic with exponent $r \geq 2$;
(2) $|H(\sigma)| \leq |P|$;
(3) $H(\sigma)$ is A-periodic and $|A| \leq |P|$;

Moreover, condition (1) holds for at least one closed variable section σ of Ω.

Let H be a P-periodic solution of Ω. Then a section σ satisfying condition (1) of the above definition is called P-*periodic* (with respect to H).

The following lemma gives an intuition about the kind of generalised equations that have periodic solutions.

LEMMA 6.2. *Let Ω be a generalised equation such that every closed section σ_i of Ω is either constant or there exists a pair of dual bases μ_i, $\Delta(\mu_i)$ such that μ_i and $\Delta(\mu_i)$ intersect but do not form a pair of matched bases, and $\sigma_i = [\alpha(\mu_i), \beta(\Delta(\mu_i))]$. Let H be a solution of Ω. Then there exists a period P such that H is P-periodic.*

PROOF. Consider a section $\sigma = [\alpha(\mu), \beta(\Delta(\mu))]$. The boundary $i_1 = \alpha(\Delta(\mu))$ intersects the base μ, since the bases μ and $\Delta(\mu)$ overlap. We μ-tie the boundary i_1, i.e. we introduce a boundary connection (i_1, μ, i_2) in such a way that H is a solution of the generalised equation obtained. It follows that $i_1 < i_2$.

Repeating this argument, we obtain a finite set of boundaries $i_1 < \cdots < i_{k+1}$, $k \geq 1$ such that $i_1, \ldots, i_k$ do and i_{k+1} does not intersect μ, there is a boundary connection (i_j, μ, i_{j+1}) for all $j = 1, \ldots, k$ and H induces a solution of the generalised equation obtained. This set of boundaries is finite, since the length of the solution H is finite.

Let $H[\alpha(\mu), i_1] = w$, $w = A^l$, where $l \geq 1$ and A is a period. Then the section σ is A-periodic. Indeed,

$$\sigma = [\alpha(\mu), i_1] \cup [i_1, i_2] \cup \cdots \cup [i_k, i_{k+1}] \cup [i_{k+1}, \beta(\Delta(\mu))].$$

Using the boundary equations, we get that

$$h[\alpha(\mu), i_1] = h[i_1, i_2] = \cdots = h[i_k, i_{k+1}] \text{ and } h[i_{k+1}, \beta(\Delta(\mu))] = h[i_k, \beta(\mu)],$$

thus $H(\sigma) \doteq A^{l \cdot (k+1)} \cdot A_1$, where $A \doteq A_1 A_2$.

Set $P = A_j$, where $|A_j| = \max_i \{|A_i| \mid \sigma_i \text{ is } A_i\text{-periodic}\}$. By definition, H is P-periodic. □

Below we introduce the notion of a periodic structure. The idea of considering periodic structures on Ω is to subdivide the set of periodic solutions into subsets so that any two solutions from the same subset have the same set of "long items", i.e. P-periodic solutions that factor through the generalised equation Ω and a periodic structure $\langle \mathcal{P}, R \rangle$ on Ω satisfy the following property:

$$h_i \in \mathcal{P} \text{ if and only if } |H_i| \geq 2|P|.$$

One can regard Lemma 6.5 below as a motivation for the definition of a periodic structure.

DEFINITION 6.3. Let Ω be a generalised equation in the standard form without boundary connections. A *periodic structure* on Ω is a pair $\langle \mathcal{P}, R \rangle$, where

(1) $\mathcal{P}$ is a non-empty set consisting of some variables h_i, some variable bases μ, and some closed sections σ from $V\Sigma$ such that the following conditions are satisfied:
    (a) if $h_i \in \mathcal{P}$ and $h_i \in \mu$, then $\mu \in \mathcal{P}$;

(b) if $\mu \in \mathcal{P}$, then $\Delta(\mu) \in \mathcal{P}$;
(c) if $\mu \in \mathcal{P}$ and $\mu \in \sigma$, then $\sigma \in \mathcal{P}$;
(d) there exists a function $\mathcal{X}$ mapping the set of closed sections from $\mathcal{P}$ into $\{-1, +1\}$ such that for every $\mu, \sigma_1, \sigma_2 \in \mathcal{P}$, the condition that $\mu \in \sigma_1$ and $\Delta(\mu) \in \sigma_2$ implies $\varepsilon(\mu) \cdot \varepsilon(\Delta(\mu)) = \mathcal{X}(\sigma_1) \cdot \mathcal{X}(\sigma_2)$;

(2) R is an equivalence relation on a certain set $\mathcal{B}$ (defined in (e)) such that condition (f) is satisfied.

(e) Notice, that for every boundary l belonging to a closed section in $\mathcal{P}$ either there exists a unique closed section $\sigma(l)$ in $\mathcal{P}$ containing l, or there exist precisely two closed sections $\sigma_{\text{left}}(l) = [i, l], \sigma_{\text{right}} = [l, j]$ in $\mathcal{P}$ containing l. The set of boundaries of the first type we denote by $\mathcal{B}_1$, and of the second type by $\mathcal{B}_2$. Put

$$\mathcal{B} = \mathcal{B}_1 \cup \{l_{\text{left}}, l_{\text{right}} \mid l \in \mathcal{B}_2\}$$

here $l_{\text{left}}, l_{\text{right}}$ are two "formal copies" of l. We will use the following agreement: for any base μ if $\alpha(\mu) \in \mathcal{B}_2$ then by $\alpha(\mu)$ we mean $\alpha(\mu)_{\text{right}}$ and, similarly, if $\beta(\mu) \in \mathcal{B}_2$ then by $\beta(\mu)$ we mean $\beta(\mu)_{\text{left}}$.

(f) If $\mu \in \mathcal{P}$ then

$$\alpha(\mu) \sim_R \alpha(\Delta(\mu)), \quad \beta(\mu) \sim_R \beta(\Delta(\mu)), \quad \text{if } \varepsilon(\mu) = \varepsilon(\Delta(\mu));$$
$$\alpha(\mu) \sim_R \beta(\Delta(\mu)), \quad \beta(\mu) \sim_R \alpha(\Delta(\mu)), \quad \text{if } \varepsilon(\mu) = -\varepsilon(\Delta(\mu)).$$

REMARK 6.4. For a given generalised equation Ω, there exists only finitely many periodic structures on Ω, and all of them can be constructed effectively. Indeed, every periodic structure $\langle \mathcal{P}, R \rangle$ is uniquely defined by the subset of items of Ω that belong to $\mathcal{P}$ and the relation $\sim_R$. Therefore, to describe all periodic structures on Ω it suffices to consider all subsets of the set of items of Ω and different relations $\sim_R$ on them.

Now we will show how to a P-periodic solution H of Ω one can associate a periodic structure $\mathcal{P}(H, P) = \langle \mathcal{P}, R \rangle$ on Ω. We define $\mathcal{P}$ as follows. A closed section σ is in $\mathcal{P}$ if and only if σ is P-periodic. A variable h_i is in $\mathcal{P}$ if and only if $h_i \in \sigma$ for some $\sigma \in \mathcal{P}$ and $|H_i| \geq 2|P|$. A variable base μ is in $\mathcal{P}$ if and only if either μ or $\Delta(\mu)$ contains an item h_i from $\mathcal{P}$.

Put $\mathcal{X}([i, j]) = \pm 1$ depending on whether in (6.1) the word Q is conjugate to P or to P^{-1}.

Now let $[i, j] \in \mathcal{P}$ and $i \leq l \leq j$. Then one can write $P \doteq P_1 P_2$ in such a way that if $\mathcal{X}([i, j]) = 1$, then the word $H[i, l]$ is the terminal subword of the word $(P^\infty)P_1$, where P^∞ is the infinite word obtained by concatenating the powers of P, and $H[l, j]$ is the initial subword of the word $P_2(P^\infty)$; and if $\mathcal{X}([i, j]) = -1$, then the word $H[i, l]$ is the terminal subword of the word $(P^{-1})^\infty P_2^{-1}$ and $H[l, j]$ is the initial subword of $P_1^{-1}(P^{-1})^\infty$. By Lemma 1.2.9 [**Ad75**], the decomposition $P \doteq P_1 P_2$ with these properties is unique; denote this decomposition by $\delta(l)$. We define the relation R as follows:

$$l_1 \sim_R l_2 \text{ if and only if } \delta(l_1) = \delta(l_2).$$

Every periodic solution H of Ω induces a periodic structure $\mathcal{P}(H, P)$ on Ω.

LEMMA 6.5. *Let H be a periodic solution of Ω. Then $\mathcal{P}(H, P)$ is a periodic structure on Ω.*

PROOF. Let $\mathcal{P}(H, P) = \langle \mathcal{P}, R \rangle$. Obviously, $\mathcal{P}$ satisfies conditions (a) and (b) from Definition 6.3.

We now prove that $\mathcal{P}$ satisfies condition (c) from Definition 6.3. Let $\mu \in \mathcal{P}$ and $\mu \in [i, j]$. There exists an item $h_k \in \mathcal{P}$ such that $h_k \in \mu$ or $h_k \in \Delta(\mu)$. If $h_k \in \mu$, then, by construction, $[i, j] \in \mathcal{P}$. If $h_k \in \Delta(\mu)$ and $\Delta(\mu) \in [i', j']$, then $[i', j'] \in \mathcal{P}$, and hence, the word $H(\Delta(\mu))$ can be written in the form $Q^{r'} Q_1$, where $Q \doteq Q_1 Q_2$ is a cyclic permutation of the word $P^{\pm 1}$ and $r' \geq 2$. Since $|H[i, j]| \geq |H(\mu)| = |H(\Delta(\mu))| \geq 2|P|$ and from Definition 6.1, it follows that $[i, j]$ is an A-periodic section, where $|A| \leq |P|$. Then $H(\mu) \doteq B^s B_1$, where B is a cyclic permutation of the word $A^{\pm 1}$, $|B| \leq |P|$, $B \doteq B_1 B_2$, and $s \geq 0$. From the equality $H(\mu)^{\varepsilon(\mu)} \doteq H(\Delta(\mu))^{\varepsilon(\Delta(\mu))}$ and Lemma 1.2.9 [**Ad75**] it follows that B is a cyclic permutation of the word $Q^{\pm 1}$. Consequently, A is a cyclic permutation of the word $P^{\pm 1}$. Therefore, $[i, j]$ is a P-periodic section of Ω with respect to H, in other words, the length of $H[i, j]$ is greater than or equal to $2|P|$ and so $[i, j] \in \mathcal{P}$.

If $\mu \in [i_1, j_1]$, $\Delta(\mu) \in [i_2, j_2]$ and $\mu \in \mathcal{P}$, then the equality $\varepsilon(\mu) \cdot \varepsilon(\Delta(\mu)) = \mathcal{X}([i_1, j_1]) \cdot \mathcal{X}([i_2, j_2])$ follows from the fact that given $A^r A_1 \doteq B^s B_1$ and $r, s \geq 2$, the word A cannot be a cyclic permutation of the word B^{-1}, hence condition (d) of Definition 6.1 holds.

Since $H(\mu)^{\varepsilon(\mu)} \doteq H(\Delta(\mu))^{\varepsilon(\Delta(\mu))}$, from Lemma 1.2.9 in [**Ad75**] it follows that condition (f) also holds for the relation R. □

REMARK 6.6. Now let us fix a non-empty periodic structure $\langle \mathcal{P}, R \rangle$ on a generalised equation Ω. Item (d) of Definition 6.3 allows us to assume (after replacing the variables $h_i, \ldots, h_{j-1}$ by $h_{j-1}^{-1}, \ldots, h_i^{-1}$ on those closed sections $[i, j] \in \mathcal{P}$ for which $\mathcal{X}([i, j]) = -1$) that $\varepsilon(\mu) = 1$ for all $\mu \in \mathcal{P}$. Therefore, we may assume that for every item h_i, the word H_i is a subword of the word P^∞.

The rest of this section is devoted to defining the group of automorphisms $\mathfrak{A}(\Omega)$. The idea is as follows. We change the set of generators h of the coordinate group $\mathbb{G}_{R(\Omega^*)}$ to $\bar{x}$ and we use the new set of generators to give an explicit description of the generating set of the group of automorphisms $\mathfrak{A}(\Omega)$.

To construct the set of generators $\bar{x}$ of $\mathbb{G}_{R(\Omega^*)}$, the following definitions are in order. We refer the reader to Section 6.2 for an example.

DEFINITION 6.7. We construct the *graph* Γ *of a periodic structure* $\langle \mathcal{P}, R \rangle$. The set of vertices $V(\Gamma)$ of the graph Γ is the set of R-equivalence classes. For a boundary k, denote by (k) the equivalence class of the relation R to which it belongs. For each variable h_k that belongs to a certain closed section from $\mathcal{P}$, we introduce an oriented edge $e \in E(\Gamma)$ from (k) to $(k+1)$, $e : (k) \to (k+1)$ and an inverse edge $e^{-1} : (k+1) \to (k)$. We label the edge e by $h(e) = h_k$ (correspondingly, $h(e^{-1}) = h_k^{-1}$). For every path $\mathfrak{p} = e_1^{\epsilon_1} \ldots e_j^{\epsilon_j}$ in the graph Γ, we denote its label by $h(\mathfrak{p})$, $h(\mathfrak{p}) = h(e_1^{\epsilon_1}) \ldots h(e_j^{\epsilon_j})$, $\epsilon_1, \ldots, \epsilon_j \in \{1, -1\}$.

The periodic structure $\langle \mathcal{P}, R \rangle$ is called *connected*, if its graph Γ is connected.

Let $\langle \mathcal{P}, R \rangle$ be an arbitrary periodic structure of a generalised equation Ω. Let $\Gamma_1, \ldots, \Gamma_r$ be the connected components of the graph Γ. The set of labels of edges of a connected component Γ_i is the union of sets of the form $\{h_j, h_{j+1}, \ldots, h_{k-1} \mid$ for some closed section $\sigma = [j, k] \in \mathcal{P}\}$. Moreover, if a base $\mu \in \mathcal{P}$ belongs to a section from $\mathcal{P}$, then its dual $\Delta(\mu)$, by condition (f) of Definition 6.3, also belongs to a section from $\mathcal{P}$. Define $\mathcal{P}_i$ to be the union of the following three sets: the

6.1. PERIODIC STRUCTURES

set of labels of edges from Γ_i that belong to $\mathcal{P}$, closed sections to which these labels belong, and bases $\mu \in \mathcal{P}$ that belong to these sections. Define R_i to be the restriction of the relation R to $\mathcal{P}_i$. We thereby obtain a connected periodic structure $\langle \mathcal{P}_i, R_i \rangle$ whose graph is Γ_i.

We further assume that the periodic structure $\langle \mathcal{P}, R \rangle$ is connected.

Let Γ be the graph of a periodic structure $\langle \mathcal{P}, R \rangle = \mathcal{P}(H, P)$ and let $\mathfrak{p} = e_1^{\epsilon_1} \ldots e_j^{\epsilon_j}$ be a path in Γ, $h(\mathfrak{p})$ be its label, $h(\mathfrak{p}) = h(e_1^{\epsilon_1}) \ldots h(e_j^{\epsilon_j})$, $\epsilon_1, \ldots, \epsilon_j \in \{1, -1\}$. To simplify the notation we write $H(\mathfrak{p})$ instead of $H(h(\mathfrak{p}))$.

LEMMA 6.8. *Let H be a P-periodic solution of a generalised equation Ω, let $\langle \mathcal{P}, R \rangle = \mathcal{P}(H, P)$ be a periodic structure on Ω and let $\mathfrak{c}$ be a cycle in the graph $\Gamma = \Gamma(\langle \mathcal{P}, R \rangle)$ at the vertex (l), $\delta(l) = P_1 P_2$. Then there exists $n \in \mathbb{Z}$ such that $H(\mathfrak{c}) = (P_2 P_1)^n$.*

PROOF. If e is an edge $v \to v'$ in the graph Γ, and $P = P_1 P_2$, $P = P_1' P_2'$ are two decompositions corresponding to the boundaries from (v) and (v') respectively. Then, obviously, $H(e) = P_2 P^{n_k} P_1'$, $n_k \in \mathbb{Z}$. The statement follows if we multiply the values $H(e)$ for all the edges e in the cycle $\mathfrak{c}$. $\square$

DEFINITION 6.9. A generalised equation Ω is called *periodised* with respect to a given periodic structure $\langle \mathcal{P}, R \rangle$, if for every two cycles $\mathfrak{c}_1$ and $\mathfrak{c}_2$ based at the same vertex in the graph Γ, there is a relation $[h(\mathfrak{c}_1), h(\mathfrak{c}_2)] = 1$ in $\mathbb{G}_{R(\Upsilon^*)}$. Note that, in particular $[h(\mathfrak{c}_1), h(\mathfrak{c}_2)] = 1$ in $\mathbb{G}_{R(\Omega^*)}$.

Let $\mathrm{Sh} = \mathrm{Sh}(\Gamma) = \{ e \in E(\Gamma) \mid h(e) \notin \mathcal{P} \}$ and $h(\mathrm{Sh}) = \{ h(e) \mid e \in \mathrm{Sh} \}$.

Let $\Gamma_0 = (V(\Gamma), \mathrm{Sh}(\Gamma))$ be the subgraph of the graph Γ having the same set of vertices as Γ and the set of edges $E(\Gamma_0) = \mathrm{Sh}$. Choose a maximal subforest T_0 in the graph Γ_0 and extend it to a maximal subforest T of the graph Γ. Since the periodic structure $\langle \mathcal{P}, R \rangle$ is connected by assumption, it follows that T is a tree. Fix an arbitrary vertex v_Γ of the graph Γ and denote by $\mathfrak{p}(v_\Gamma, v)$ the (unique) path in T from v_Γ to v. For every edge $e : v \to v'$ not lying in T, we introduce a cycle $\mathfrak{c}_e = \mathfrak{p}(v_\Gamma, v) e (\mathfrak{p}(v_\Gamma, v'))^{-1}$. Then the fundamental group $\pi_1(\Gamma, v_\Gamma)$ is generated by the cycles $\mathfrak{c}_e$ (see, for example, the proof of Proposition III.2.1, [**LS77**]). This and the decidability of the universal theory of $\mathbb{G}$ imply that the property of a generalised equation "to be periodised with respect to a given periodic structure" is algorithmically decidable. Indeed, it suffices to check if the following universal formula (quasi-identity) holds in $\mathbb{G}$ (for every pair of cycles $\mathfrak{c}_{e_1}, \mathfrak{c}_{e_2}$):

$$\forall H_1, \ldots, H_\rho \left(\left(\bigwedge (\Upsilon^*(H) = 1) \right) \to ([H(\mathfrak{c}_{e_1}), H(\mathfrak{c}_{e_2})] = 1) \right).$$

Furthermore, the set of elements

(6.2) $$\{ h(e) \mid e \in T \} \cup \{ h(\mathfrak{c}_e) \mid e \notin T \}$$

forms a basis of the free group generated by

$$\{ h_k \mid h_k \in \sigma, \sigma \in \mathcal{P} \}.$$

If $\mu \in \mathcal{P}$, then $(\beta(\mu)) = (\beta(\Delta(\mu)))$, $(\alpha(\mu)) = (\alpha(\Delta(\mu)))$ by property (f) from Definition 6.3 and, consequently, the word $h(\mu) h(\Delta(\mu))^{-1}$ is the label of a cycle $\mathfrak{c}_\mu'$ from $\pi_1(\Gamma, (\alpha(\mu)))$. Let

$$\mathfrak{c}_\mu = \mathfrak{p}(v_\Gamma, (\alpha(\mu))) \mathfrak{c}_\mu' \mathfrak{p}(v_\Gamma, (\alpha(\mu)))^{-1}.$$

Then
(6.3) $$h(\mathfrak{c}_\mu) = uh(\mu)h(\Delta(\mu))^{-1}u^{-1},$$
where u is the label of the path $\mathfrak{p}(v_\Gamma, (\alpha(\mu)))$. Since $\mathfrak{c}_\mu \in \pi_1(\Gamma, v_\Gamma)$, it follows that $\mathfrak{c}_\mu = b_\mu(\{\mathfrak{c}_e \mid e \notin T\})$, where b_μ is a certain word in the indicated generators that can be constructed effectively (see Proposition III.2.1, [**LS77**]).

Let $\tilde{b}_\mu$ denote the image of the word b_μ in the abelianisation of $\pi_1(\Gamma, v_\Gamma)$. Denote by $\widetilde{Z}$ the free abelian group consisting of formal linear combinations $\sum_{e \notin T} n_e \tilde{\mathfrak{c}}_e$, $n_e \in \mathbb{Z}$, and by $\widetilde{B}$ its subgroup generated by the elements $\tilde{b}_\mu$, $\mu \in \mathcal{P}$ and the elements $\tilde{\mathfrak{c}}_e$, $e \notin T$, $e \in \text{Sh}$.

By the classification theorem of finitely generated abelian groups, one can effectively construct a basis $\{\widetilde{C}^{(1)}, \widetilde{C}^{(2)}\}$ of $\widetilde{Z}$ such that
(6.4) $$\widetilde{Z} = \widetilde{Z}_1 \oplus \widetilde{Z}_2, \ \widetilde{B} \subseteq \widetilde{Z}_1, \ [\widetilde{Z}_1 : \widetilde{B}] < \infty,$$
where $\widetilde{C}^{(1)}$ is a basis of $\widetilde{Z}_1$ and $\widetilde{C}^{(2)}$ is a basis of $\widetilde{Z}_2$.

By Proposition I.4.4 in [**LS77**], one can effectively construct a basis $C^{(1)}$, $C^{(2)}$ of the free (non-abelian) group $\pi_1(\Gamma, v_\Gamma)$ such that $\widetilde{C}^{(1)}$, $\widetilde{C}^{(2)}$ are the natural images of the elements $C^{(1)}$, $C^{(2)}$ in $\widetilde{Z}$.

REMARK 6.10. Notice that any equation in $\widetilde{Z}$ of the form
$$\tilde{\mathfrak{c}} = \sum_{\tilde{\mathfrak{c}}_i \in \widetilde{Z}} n_i \tilde{\mathfrak{c}}_i$$
lifts to an equation in $\pi_1(\Gamma, v_\Gamma)$ of the form
$$\mathfrak{c} = \prod_{\mathfrak{c}_i \in \pi_1(\Gamma, v_\Gamma)} \mathfrak{c}_i^{n_i} V,$$
where V is an element of the derived subgroup of $\pi_1(\Gamma, v_\Gamma)$. If the generalised equation is periodised, for any two cycles $\mathfrak{c}'_1, \mathfrak{c}'_2 \in \pi_1(\Gamma, v_\Gamma)$, we have that $[h(\mathfrak{c}'_1), h(\mathfrak{c}'_2)] = 1$ in $\mathbb{G}_{R(\Omega^*)}$. Hence, $h(\mathfrak{c}) = \prod_{\mathfrak{c}_i \in \pi_1(\Gamma, v_\Gamma)} h(\mathfrak{c}_i)^{n_i}$ in $\mathbb{G}_{R(\Omega^*)}$.

LEMMA 6.11. *Let Ω be a periodised generalised equation. Then the basis $\widetilde{C}^{(1)}$ can be chosen in such a way that for every $\mathfrak{c} \in C^{(1)}$ either $\mathfrak{c} = \mathfrak{c}_e$, where $e \notin T$, $e \in \text{Sh}$, or for any solution H we have $H(\mathfrak{c}) = 1$.*

PROOF. The set $\{\tilde{\mathfrak{c}}_e \mid e \notin T, e \in \text{Sh}\}$ is a subset of the set of generators of $\widetilde{Z}$ contained in $\widetilde{Z}_1$. Thus, this set can be extended to a basis of $\widetilde{Z}_1$. Since, by (6.4), $[\widetilde{Z}_1 : \widetilde{B}] < \infty$, for every $\tilde{\mathfrak{c}} \in \widetilde{Z}_1$ there exists $n_\mathfrak{c} \in \mathbb{N}$ such that
$$n_\mathfrak{c} \tilde{\mathfrak{c}} = \sum_{e \notin T, e \in \text{Sh}} n_e \tilde{\mathfrak{c}}_e + \sum_{\mu \in \mathcal{P}} n_\mu \tilde{b}_\mu.$$
It follows that the set $\{\tilde{\mathfrak{c}}_1, \ldots, \tilde{\mathfrak{c}}_k\}$ which completes the set $\{\tilde{\mathfrak{c}}_e \mid e \notin T, e \in \text{Sh}\}$ to a basis of $\widetilde{Z}_1$, can be chosen so that the following equality holds:
$$n_{\mathfrak{c}_i} \tilde{\mathfrak{c}}_i = \sum_{\mu \in \mathcal{P}} n_\mu \tilde{b}_\mu.$$
Hence, by Remark 6.10, for any solution H we have $H(\mathfrak{c}_i)^{n_{\mathfrak{c}_i}} = \prod_{\mu \in \mathcal{P}} H(b_\mu)^{n_\mu} = 1$. Since $\mathbb{G}$ is torsion-free, we have $H(\mathfrak{c}_i) = 1$. □

6.1. PERIODIC STRUCTURES

Let $\{e_1, \ldots, e_m\}$ be the set of edges of $T \setminus T_0$. Since T_0 is the spanning forest of the graph Γ_0, it follows that $h(e_1), \ldots, h(e_m) \notin h(\mathrm{Sh})$; in particular, $h(e_1), \ldots, h(e_m) \in \mathcal{P}$.

Let $F(\Omega)$ be the free group generated by the variables of Ω. Consider in the group $F(\Omega)$ a new set of generators $\bar{x}$ defined below (we prove that the set $\bar{x}$ is in fact a set of generators of $F(\Omega)$ in part (1) of Lemma 6.14).

Let v_i be the origin of the edge e_i. We introduce new variables

(6.5) $\quad \bar{u}^{(i)} = \{u_{ie} \mid e \notin T, \, e \in \mathrm{Sh}\}, \quad \bar{z}^{(i)} = \{z_{ie} \mid e \notin T, e \in \mathrm{Sh}\}, \text{ for } 1 \leq i \leq m,$

as follows

(6.6) $\quad u_{ie} = h(\mathfrak{p}(v_\Gamma, v_i))^{-1} h(\mathfrak{c}_e) h(\mathfrak{p}(v_\Gamma, v_i)), \quad z_{ie} = h(e_i)^{-1} u_{ie} h(e_i).$

We denote by $\bar{u}$ the union $\bigcup_{i=1}^{m} \bar{u}^{(i)}$ and by $\bar{z}$ the union $\bigcup_{i=1}^{m} \bar{z}^{(i)}$. Denote by $\bar{t}$ the family of variables that do not belong to closed sections from $\mathcal{P}$. Let

$$\bar{x} = \bar{t} \cup \{h(e) \mid e \in T_0\} \cup \bar{u} \cup \bar{z} \cup \{h(e_1), \ldots, h(e_m)\} \cup h(C^{(1)}) \cup h(C^{(2)})$$

REMARK 6.12. Note that without loss of generality we may assume that v_Γ corresponds to the beginning of the period P. Indeed, it follows from the definition of the periodic structure that for any cyclic permutation $P' = P_2 P_1$ of P, we have $\mathcal{P}(H, P) = \mathcal{P}(H, P')$.

LEMMA 6.13. *Let Ω be a generalised equation periodised with respect to a periodic structure $\langle \mathcal{P}, R \rangle$. Then for any cycle $\mathfrak{c}_{e_0}$ such that $h(e_0) \notin \mathcal{P}$ and for any solution H of Ω periodic with respect to a period P, such that $\mathcal{P}(H, P) = \langle \mathcal{P}, R \rangle$ one has $H(\mathfrak{c}_{e_0}) = P^n$, where $|n| \leq 2\rho$. In particular, one can choose a basis $C^{(1)}$ in such a way that for any $\mathfrak{c} \in C^{(1)}$ one has $H(\mathfrak{c}) = P^n$, where $|n| \leq 2\rho$.*

PROOF. Let $\mathfrak{c}_{e_0}$ be a cycle such that the edges of $\mathfrak{c}_{e_0}$ are labelled by variables h_k, $h_k \notin \mathcal{P}$. Observe that $e_0 = \mathfrak{p}_1 \mathfrak{c}_{e_0} \mathfrak{p}_2$, where $\mathfrak{p}_1$ and $\mathfrak{p}_2$ are paths in the tree T. Since $e_0 \in \Gamma_0$, it follows that the origin and the terminus of the edge e_0 lie in the same connected component of the graph Γ_0 and, consequently, are connected by a path $\mathfrak{s}$ in the forest T_0. Furthermore, $\mathfrak{p}_1$ and $\mathfrak{s}\mathfrak{p}_2^{-1}$ are paths in the tree T connecting the same vertices; therefore, $\mathfrak{p}_1 = \mathfrak{s}\mathfrak{p}_2^{-1}$. Hence, $\mathfrak{c}_{e_0} = \mathfrak{p}_2 \mathfrak{c}'_{e_0} \mathfrak{p}_2^{-1}$, where $\mathfrak{c}'_{e_0}$ is a certain cycle based at the vertex v'_Γ in the graph Γ_0.

From the equality $H(\mathfrak{c}_{e_0}) = H(\mathfrak{p}_2) H(\mathfrak{c}'_{e_0}) H(\mathfrak{p}_2)^{-1}$ and by Lemma 6.8, we get that $P^{n_{e_0}} = P^{n_2} P_1 (P_2 P_1)^{n'_{e_0}} P_1^{-1} P^{-n_2}$, where $\delta(v'_\Gamma) = P_1 P_2$, $n_{e_0} = \exp(H(\mathfrak{c}_{e_0}))$, $n'_{e_0} = \exp(H(\mathfrak{c}'_{e_0}))$, $n_2 = \exp(H(\mathfrak{p}_2))$. Hence $n_{e_0} = n'_{e_0}$ and thus $|H(\mathfrak{c}_{e_0})| = |H(\mathfrak{c}'_{e_0})|$. From the construction of $\mathcal{P}(H, P)$, it follows that the inequality $|H_k| \leq 2|P|$ holds for every item $h_k \notin \mathcal{P}$. Since the cycle $\mathfrak{c}'_{e_0}$ is simple, we have that $|H(\mathfrak{c}_{e_0})| = |H(\mathfrak{c}'_{e_0})| \leq 2\rho |P|$.

In particular, one has that $|\exp(H(\mathfrak{c}_e))| \leq 2\rho$ for every $e \notin T$, $e \in \mathrm{Sh}$. By Lemma 6.11 one can choose a basis $C^{(1)}$ such that for every $\mathfrak{c} \in C^{(1)}$ either $\mathfrak{c} = \mathfrak{c}_e$ and $|\exp(H(\mathfrak{c}_e))| \leq 2\rho$, or $H(\mathfrak{c}) = 1$. $\square$

The following lemma describes a generating set and the group of automorphisms $\mathfrak{A}(\Omega)$ of the coordinate group $\mathbb{G}_{R(\Upsilon^*)}$.

LEMMA 6.14. *Let Ω be a generalised equation periodised with respect to a periodic structure $\langle \mathcal{P}, R \rangle$. Then the following statements hold.*

(1) The system Υ^* is equivalent to the union of the following two systems of equations:
$$\begin{cases} u_{ie}^{h(e_i)} = z_{ie}, & \text{where } e \in T, e \in \text{Sh}; 1 \leq i \leq m \\ [u_{ie_1}, u_{ie_2}] = 1, & \text{where } e_j \in T, e_j \in \text{Sh}, j = 1, 2; 1 \leq i \leq m \\ [h(\mathfrak{c}_1), h(\mathfrak{c}_2)] = 1, & \text{where } \mathfrak{c}_1, \mathfrak{c}_2 \in C^{(1)} \cup C^{(2)} \end{cases}$$

and a system:
$$\Psi\left(\{h(e) \mid e \in T, e \in \text{Sh}\}, h(C^{(1)}), \bar{t}, \bar{u}, \bar{z}, \mathcal{A}\right) = 1,$$

such that neither $h(e_i)$, $1 \leq i \leq m$, nor $h(C^{(2)})$ occurs in Ψ.

(2) If $\mathfrak{c}$ is a cycle based at the origin of e_i, then the transformation $h(e_i) \to h(\mathfrak{c})h(e_i)$ which is identical on all the other elements from $\mathcal{A} \cup \bar{x}$, extends to a $\mathbb{G}$-automorphism of $\mathbb{G}_{R(\Upsilon^*)}$.

(3) If $\mathfrak{c} \in C^{(2)}$ and $\mathfrak{c}' \in C^{(1)} \cup C^{(2)}$, $\mathfrak{c}' \neq \mathfrak{c}$, then the transformation defined by $h(\mathfrak{c}) \to h(\mathfrak{c}')h(\mathfrak{c})$, which is identical on all the other elements from $\mathcal{A} \cup \bar{x}$, extends to a $\mathbb{G}$-automorphism of $\mathbb{G}_{R(\Upsilon^*)}$.

PROOF. We first prove (1). It is easy to check that (in the above notation):
$$\{h_1, \ldots, h_\rho\} = \{h(e) \mid h(e) \in \sigma, \sigma \notin \mathcal{P}\} \cup$$
$$\{h(e) \mid e \in T_0\} \cup \{h(e) \mid e \notin T_0, e \in \text{Sh}\} \cup$$
$$\{h(e) \mid e \in T \setminus T_0\} \cup \{h(e) \mid e \notin T, e \notin \text{Sh}\}.$$

From Equation (6.2) and the discussion above (recall that the set $C^{(1)} \cup C^{(2)}$ generates $\pi_1(\Gamma)$) it follows that the set
$$\{\bar{t} \cup \{h(e) \mid e \in T_0\} \cup \{h(e_1), \ldots, h(e_m)\} \cup h(C^{(1)}) \cup h(C^{(2)})\}$$
is a basis of the free group $F(\Omega) = \langle h_1, \ldots, h_\rho \rangle$, hence the set $\bar{x}$ is a generating set for $F(\Omega)$:

(6.7) $\quad \langle h_1, \ldots, h_\rho \rangle = \left\langle \begin{array}{c} \bar{t} \cup \{h(e) \mid e \in T_0\} \cup \bar{u} \cup \bar{z} \\ \cup \{h(e_1), \ldots, h(e_m)\} \cup h(C^{(1)}) \cup h(C^{(2)}) \end{array} \right\rangle = \langle \bar{x} \rangle.$

We now rewrite the equations from Υ in terms of the new set of generators $\bar{x}$.

We first consider the relations induced by bases which do not belong to $\mathcal{P}$, i.e. basic equations of the form $h(\mu) = h(\Delta(\mu))$, where μ is a variable base, $\mu \notin \mathcal{P}$. Since $\mu \notin \mathcal{P}$, it follows by Definition 6.3 that $\Delta(\mu) \notin \mathcal{P}$ and no item from μ or $\Delta(\mu)$ belongs to $\mathcal{P}$. It now follows by construction of the generating set $\bar{x}$ that all the items h_k which appear in these relations belong to the set
$$\bar{t} \cup \{h(e) \mid e \in T_0\} \cup \{h(e) \mid e \notin T_0, e \in \text{Sh}\} = \bar{t} \cup M_1 \cup M_2.$$

The elements of the sets $\bar{t}$ and M_1 are generators in both basis. We now study how elements of M_2 rewrite in the new basis. Let $h(e) = h_k$, $e \notin T_0$, $e \in \text{Sh}$. We have $e = \mathfrak{sp}_2^{-1} \mathfrak{c}_e \mathfrak{p}_2$ and

(6.8) $\quad h_k = h(\mathfrak{s}) h(\mathfrak{p}_2)^{-1} h(\mathfrak{c}_e) h(\mathfrak{p}_2),$

where $\mathfrak{s}$ is a path in T_0 and $\mathfrak{p}_2$ is a path in T (see proof of Lemma 6.13). The variables $h(e_i)$, $1 \leq i \leq m$ can occur in the right-hand side of Equation (6.8) (written in the basis $\bar{x}$) only in $h(\mathfrak{p}_2)^{\pm 1}$ and at most once. Moreover, the sign of this occurrence (if it exists) depends only on the orientation of the edge e_i with respect to the root v_Γ of the tree T. If $\mathfrak{p}_2 = \mathfrak{p}'_2 e_i^{\pm 1} \mathfrak{p}''_2$, then all the occurrences of the variable $h(e_i)$ in the words h_k written in the basis $\bar{x}$, with $h_k \notin \mathcal{P}$, are contained in the subwords of the form $h(e_i)^{\mp 1} h((\mathfrak{p}_2')^{-1} \mathfrak{c}_e \mathfrak{p}_2') h(e_i)^{\pm 1}$, i.e. in the subwords of the form $h(e_i)^{\mp 1} h(\mathfrak{c}) h(e_i)^{\pm 1}$, where $\mathfrak{c}$ is a certain cycle in the graph Γ based at the origin of the edge $e_i^{\pm 1}$. So the variable h_k rewrites as a word in the generators $\bar{u}$, $\bar{z}$, $\{h(e) \mid e \in T_0\}$.

Summarising, the basic equations corresponding to variable bases which do not belong to $\mathcal{P}$ rewrite in the new basis as words in $\bar{t}$, $\{h(e) \mid e \in T_0\}$, $\bar{u}$ and $\bar{z}$.

In the new basis, the relations of the form $h(\mu) = h(\Delta(\mu))$, where $\mu \in \mathcal{P}$, are, modulo commutators, words in $h(C^{(1)})$ (recall that $\widetilde{Z}$ was chosen to contain all the $\widetilde{b}_\mu$ for all $\mu \in \mathcal{P}$), see Remark 6.10.

Since Ω is periodised with respect to $\langle \mathcal{P}, R \rangle$, we have

(6.9) $\qquad [u_{ie_1}, u_{ie_2}] = 1$ and $[h(\mathfrak{c}_1), h(\mathfrak{c}_2)] = 1$, $\mathfrak{c}_1, \mathfrak{c}_2 \in C^{(1)} \cup C^{(2)}$.

The set $\bar{x}$ is a generating set, but not a basis of $F(\Omega) = \langle \bar{x} \mid K \rangle$. From the definition of $\bar{z}$ and $\bar{u}$, it follows that the set of relations K is the disjoint union of the set
$$\{u_{ie}^{h(e_i)} = z_{ie} \mid e \notin T, e \in \mathrm{Sh}; 1 \leq i \leq m\}$$
and a system
$$\Psi_0 \left(\{h(e) \mid e \in T, h(e) \notin \mathcal{P}\}, h(C^{(1)}), \bar{u}, \bar{z}, \mathcal{A} \right) = 1.$$

We conclude that the system Υ^* is equivalent to the union of the following two systems of equations in the new variables, a system
$$\mathcal{O} = \begin{cases} u_{ie}^{h(e_i)} = z_{ie}, & \text{where } e \notin T, e \in \mathrm{Sh}; 1 \leq i \leq m \\ [u_{ie_1}, u_{ie_2}] = 1, & \text{where } e_j \notin T, e_j \in \mathrm{Sh}, j = 1,2; 1 \leq i \leq m \\ [h(\mathfrak{c}_1), h(\mathfrak{c}_2)] = 1, & \text{where } \mathfrak{c}_1, \mathfrak{c}_2 \in C^{(1)} \cup C^{(2)} \end{cases}$$

and a system (defined by the equations from Υ and Ψ_0):
$$\Psi \left(\{h(e) \mid e \in T, h(e) \notin \mathcal{P}\}, h(C^{(1)}), \bar{t}, \bar{u}, \bar{z}, \mathcal{A} \right) = 1,$$

such that neither $h(e_i)$, $1 \leq i \leq m$, nor $h(C^{(2)})$ occurs in Ψ.

The transformations from statements (2) and (3) extend to an automorphism φ of $\mathbb{G}_{R(\Upsilon^*)}$. Indeed, by the universal property of the quotient, the following diagram commutes

$$\begin{array}{ccc} \mathbb{G}[\bar{x}] & \xrightarrow{\varphi} & \mathbb{G}[\bar{x}] \\ \downarrow & & \downarrow \\ \mathbb{G}_{R(\Psi \cup \mathcal{O})} & \dashrightarrow{\tilde{\varphi}} & \mathbb{G}_{R(\Psi \cup \mathcal{O} \cup \varphi(\mathcal{O}))} \end{array}$$

It is easy to check that $\varphi(\Psi) = \Psi$ and that $\varphi(\mathcal{O}) \subseteq R(\Psi \cup \mathcal{O})$. Therefore, since by statement (1) of the lemma the system $\Psi \cup \mathcal{O}$ is equivalent to Υ^*, we get that $\tilde{\varphi}$ is an automorphism of $\mathbb{G}_{R(\Upsilon^*)}$. $\square$

DEFINITION 6.15. Let $\Omega = \langle \Upsilon, \Re_\Upsilon \rangle$ be a generalised equation and let $\langle \mathcal{P}, R \rangle$ be a connected periodic structure on Ω. We say that the generalised equation Ω is *strongly singular with respect to the periodic structure* $\langle \mathcal{P}, R \rangle$ if one of the following conditions holds.

(a) The generalised equation Ω is not periodised with respect to the periodic structure $\langle \mathcal{P}, R \rangle$.

(b) The generalised equation Ω is periodised with respect to the periodic structure $\langle \mathcal{P}, R \rangle$ and there exists an automorphism φ of the coordinate group $\mathbb{G}_{R(\Upsilon^*)}$ of the form described in parts (2) or (3) of Lemma 6.14, such that φ does not induce an automorphism of $\mathbb{G}_{R(\Omega^*)}$.

We say that the generalised equation Ω is *singular with respect to the periodic structure* $\langle \mathcal{P}, R \rangle$ if Ω is not strongly singular with respect to the periodic structure $\langle \mathcal{P}, R \rangle$ and one of the following conditions holds

(a) The set $C^{(2)}$ has more than one element.
(b) The set $C^{(2)}$ has exactly one element, and (in the above notation) there exists a cycle $\mathfrak{c}_{e_0} \in \langle C^{(1)} \rangle$, $h(e_0) \notin \mathcal{P}$ such that $h(\mathfrak{c}_{e_0}) \neq 1$ in $\mathbb{G}_{R(\Omega^*)}$.

Otherwise, we say that Ω is *regular with respect to the periodic structure* $\langle \mathcal{P}, R \rangle$. In particular if Ω is singular or regular with respect to the periodic structure $\langle \mathcal{P}, R \rangle$ then Ω is periodised.

When no confusion arises, instead of saying that Ω is (strongly) singular (or regular) with respect to the periodic structure $\langle \mathcal{P}, R \rangle$ we say that the periodic structure $\langle \mathcal{P}, R \rangle$ is (*strongly*) *singular* (or *regular*).

DEFINITION 6.16. Let $\Omega = \langle \Upsilon, \Re_\Upsilon \rangle$ be a generalised equation and let $\langle \mathcal{P}, R \rangle$ be a periodic structure on Ω. If Ω is strongly singular with respect to $\langle \mathcal{P}, R \rangle$, then we define the group $\mathfrak{A}(\Omega)$ of automorphisms of $\mathbb{G}_{R(\Omega^*)}$ to be trivial.

Otherwise, i.e. if $\langle \mathcal{P}, R \rangle$ is singular or regular, we set $\mathfrak{A}(\Omega)$ to be the group of automorphisms of $\mathbb{G}_{R(\Omega^*)}$ generated by the automorphisms induced by the automorphisms of $\mathbb{G}_{R(\Upsilon^*)}$ described in statements (2) and (3) of Lemma 6.14. Note that by definition of singular and regular periodic structures, every such automorphism of $\mathbb{G}_{R(\Upsilon^*)}$ induces an automorphism of $\mathbb{G}_{R(\Omega^*)}$ and, therefore the group $\mathfrak{A}(\Omega)$ is finitely generated.

6.2. Example

In this section we give an example of a generalised equation Ω, a periodic structure on the generalised equation Ω and some of the constructions used in Section 6.1.

Let Ω be a generalised equation shown on Figure 12 (the one to which the entire transformation is applied). Let $\langle \mathcal{P}, R \rangle$ be a periodic structure on Ω defined as follows. Set $h_1, h_3, h_4, h_5, h_7 \in \mathcal{P}$. From part (a) of Definition 6.3 it follows that

$$\lambda_1, \lambda_3, \nu, \Delta(\lambda_1), \Delta(\lambda_3), \Delta(\nu) \in \mathcal{P}.$$

From part (c) of Definition 6.3 we get that then $\sigma = [1, 8] \in \mathcal{P}$. The set $\mathcal{B}$ in part (e) of Definition 6.3 is defined to be $\{1, \ldots, 8\}$. For the bases $\lambda_1, \lambda_3, \nu$ and their duals, we have

$$[1 = \alpha(\lambda_1)] \sim_R [\alpha(\Delta(\lambda_1)) = 5 = \beta(\mu)] \sim_R$$
$$[\beta(\Delta(\mu)) = 8 = \beta(\Delta(\lambda_3))] \sim_R [\beta(\lambda_3) = 4];$$
$$[2 = \beta(\lambda_1)] \sim_R [\beta(\Delta(\lambda_1)) = 6]; \quad [3 = \alpha(\lambda_3)] \sim_R [\alpha(\Delta(\lambda_3)) = 7].$$

Therefore, there are three R-equivalence classes: $\{1, 4, 5, 8\}, \{2, 6\}, \{3, 7\}$.

Suppose that Ω is not strongly singular with respect to the periodic structure $\langle \mathcal{P}, R \rangle$. Note that, using the decidability of the universal theory of $\mathbb{G}$, one could check if $\langle \mathcal{P}, R \rangle$ is strongly singular or not.

6.2. EXAMPLE

We construct the graph Γ of the periodic structure $\langle \mathcal{P}, R \rangle$ as in Definition 6.7. The graph Γ is shown on Figure 1. The edges e_i of the graph Γ are labelled by the corresponding items h_i. In this example the set $\mathrm{Sh}(\Gamma)$ equals $\{e_2, e_6\}$. We take the subgraph Γ_0 of Γ whose edges are elements of Sh and choose a maximal subtree T_0 of Γ_0 as shown on Figure 1. We extend the tree T_0 to a maximal subtree T of Γ, which is also shown on Figure 1.

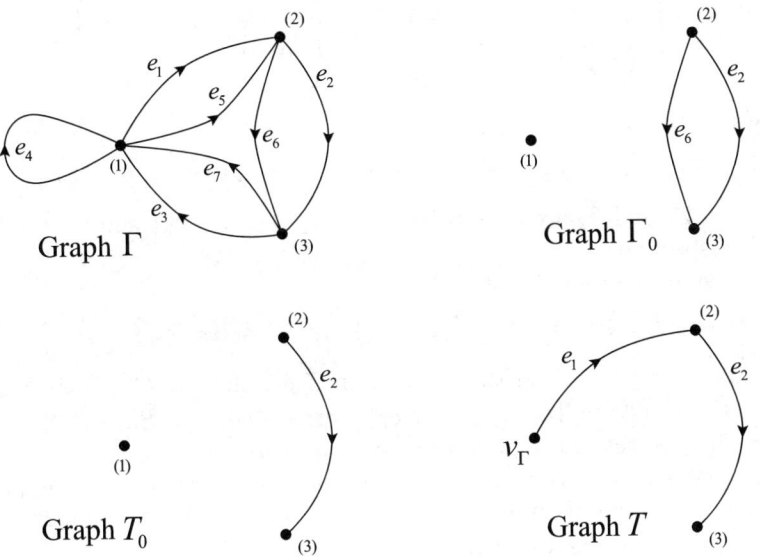

FIGURE 1. The graphs Γ, Γ_0, T_0 and T.

We fix a vertex v_Γ in Γ. Then the fundamental group $\pi_1(\Gamma, v_\Gamma)$ of Γ with respect to v_Γ is generated by the cycles $\mathfrak{c}_{e_3}, \mathfrak{c}_{e_4}, \mathfrak{c}_{e_5}, \mathfrak{c}_{e_6}, \mathfrak{c}_{e_7}$:

$$\pi_1(\Gamma, v_\Gamma) = \langle \mathfrak{c}_{e_3}, \mathfrak{c}_{e_4}, \mathfrak{c}_{e_5}, \mathfrak{c}_{e_6}, \mathfrak{c}_{e_7} \rangle, \text{ where}$$

$$\mathfrak{c}_{e_3} = e_1 e_2 e_3; \qquad \mathfrak{c}_{e_5} = e_5 e_1^{-1}; \qquad \mathfrak{c}_{e_4} = e_4;$$
$$\mathfrak{c}_{e_6} = e_1 e_6 e_2^{-1} e_1^{-1}; \qquad \mathfrak{c}_{e_7} = e_1 e_2 e_7.$$

Since, by definition, for any base $\mu \in \mathcal{P}$ one has that $\beta(\mu) \sim_R \beta(\Delta(\mu))$, the word $h(\mu)h(\Delta(\mu))^{-1}$ labels a cycle $\mathfrak{c}_\mu$ in the graph Γ. For the bases $\lambda_1, \lambda_3, \nu \in \mathcal{P}$, we construct the corresponding cycles $\mathfrak{c}_{\lambda_1}, \mathfrak{c}_{\lambda_3}, \mathfrak{c}_\nu$ as in Equation (6.3):

$$\mathfrak{c}_{\lambda_1} = e_1 e_5^{-1}, \qquad \mathfrak{c}_{\lambda_3} = e_1 e_2 e_3 e_7^{-1} e_2^{-1} e_1^{-1}, \qquad \mathfrak{c}_\nu = e_1 e_2 e_3 e_4 e_7^{-1} e_6^{-1} e_5^{-1} e_4^{-1}.$$

Since the cycles $\mathfrak{c}_{\lambda_1}, \mathfrak{c}_{\lambda_3}, \mathfrak{c}_\nu$ belong to $\pi_1(\Gamma, v_\Gamma)$, these cycles can be written as words $b_{\lambda_1}, b_{\lambda_3}, b_\nu$ in the generators $\mathfrak{c}_{e_i}$, $i = 3, 4, 5, 6, 7$ as follows:

$$b_{\lambda_1} = \mathfrak{c}_{e_5}^{-1}, \qquad b_{\lambda_3} = \mathfrak{c}_{e_3} \mathfrak{c}_{e_7}^{-1}, \qquad b_\nu = \mathfrak{c}_{e_3} \mathfrak{c}_{e_4} \mathfrak{c}_{e_7}^{-1} \mathfrak{c}_{e_6}^{-1} \mathfrak{c}_{e_5}^{-1} \mathfrak{c}_{e_4}^{-1}.$$

Let $\widetilde{Z}$ be the free abelian group generated by the images of the cycles $\mathfrak{c}_{e_i}$, $i = 3, 4, 5, 6, 7$ in the abelianisation of $\pi_1(\Gamma, v_\Gamma)$:

$$\widetilde{Z} = \langle \widetilde{\mathfrak{c}}_{e_3} \rangle \oplus \langle \widetilde{\mathfrak{c}}_{e_4} \rangle \oplus \langle \widetilde{\mathfrak{c}}_{e_5} \rangle \oplus \langle \widetilde{\mathfrak{c}}_{e_6} \rangle \oplus \langle \widetilde{\mathfrak{c}}_{e_7} \rangle = \langle e_1 + e_2 + e_3, \; e_4, \; e_5 - e_1, \; e_6 - e_2, \; e_1 + e_2 + e_7 \rangle.$$

Consider the subgroup $\widetilde{B}$ of $\widetilde{Z}$ generated by the images of $b_{\lambda_1}, b_{\lambda_3}, b_\nu$ and $\mathfrak{c}_{e_6}$ in the abelianisation of $\pi_1(\Gamma, v_\Gamma)$:

$$\widetilde{B} = \langle \widetilde{b}_{\lambda_1} \rangle \oplus \langle \widetilde{b}_{\lambda_3} \rangle \oplus \langle \widetilde{b}_\nu \rangle \oplus \langle \widetilde{\mathfrak{c}}_{e_6} \rangle = \langle e_1 - e_5, \; e_3 - e_7, \; e_1 + e_2 + e_3 - e_7 - e_6 - e_5, \; e_6 - e_5 \rangle.$$

By the classification theorem of finitely generated abelian groups, there exist free abelian groups $\widetilde{Z}_1$ and $\widetilde{Z}_2$ such that $\widetilde{Z} = \widetilde{Z}_1 \oplus \widetilde{Z}_2$ and $[\widetilde{Z}_1 : \widetilde{B}] < \infty$. To construct this decomposition (see Equation (6.4)), we construct a matrix with rows corresponding to the decomposition of the generators of $\widetilde{B}$ in terms of the generators of $\widetilde{Z}$ and then apply the elementary transformations:

$$\begin{pmatrix} 0 & 0 & -1 & 0 & 0 \\ 1 & 0 & 0 & 0 & -1 \\ 1 & 0 & -1 & -1 & -1 \\ 0 & 0 & 0 & 1 & 0 \end{pmatrix} \rightsquigarrow \begin{pmatrix} 0 & 0 & -1 & 0 & 0 \\ 1 & 0 & 0 & 0 & 0 \\ 0 & 0 & 0 & 0 & 0 \\ 0 & 0 & 0 & 1 & 0 \end{pmatrix}$$

It follows that

$$\widetilde{B} = \widetilde{Z}_1, \quad \widetilde{Z}_1 = \langle e_3 - e_7, e_5 - e_1, e_6 - e_2 \rangle, \quad \widetilde{Z}_2 = \langle e_4, e_1 + e_2 + e_7 \rangle.$$

The basis of $\widetilde{Z}_1$ and $\widetilde{Z}_2$ are $\widetilde{C}^{(1)} = \{e_3 - e_7, e_5 - e_1, e_6 - e_2\}$ and $\widetilde{C}^{(2)} = \{e_4, e_1 + e_2 + e_7\}$, respectively.

It is easy to see that the sets

$$C^{(1)} = \{\mathfrak{c}_{e_3}\mathfrak{c}_{e_7}^{-1}, \mathfrak{c}_{e_5}, \mathfrak{c}_{e_6}\}, \quad C^{(2)} = \{\mathfrak{c}_{e_4}, \mathfrak{c}_{e_7}\}$$

form a base of $\pi_1(\Gamma, v_\Gamma)$ and their images in $\widetilde{Z}$ are the sets $\widetilde{C}^{(1)} = \{e_3 - e_7, e_5 - e_1, e_6 - e_2\}$ and $\widetilde{C}^{(2)} = \{e_4, e_1 + e_2 + e_7\}$, correspondingly. Since $|C^{(2)}| = 2$, Ω is singular with respect to the periodic structure $\langle \mathcal{P}, R \rangle$.

Note that for every solution H of Ω we have $H(C^{(1)}) = 1$. Indeed, since, by assumption, Ω is periodised with respect to $\langle \mathcal{P}, R \rangle$, the following equalities hold in the coordinate group $\mathbb{G}_{R(\Omega^*)}$:

$$\mathfrak{c}_{e_3}\mathfrak{c}_{e_7}^{-1} = b_{\lambda_3}, \quad \mathfrak{c}_{e_5} = b_{\lambda_1}^{-1}, \quad \mathfrak{c}_{e_6} = b_{\lambda_3}^{-1} b_\nu b_{\lambda_1}^{-1}.$$

The statement follows, since for any cycle $\mathfrak{c}_\mu$, one has $H(\mathfrak{c}_\mu) = 1$ for any solution H.

We construct the set of generators $\bar{x}$ of $\mathbb{G}_{R(\Omega^*)}$ used in Lemma 6.14. The set $\bar{x}$ is a union of $\bar{t} = \{h_8\}$, $\{h_2\}$ (since $e_2 \in T_0$), $\{u_{1e_6} = h(e_1 e_6 e_2^{-1} e_1^{-1})\}$, $\{z_{1e_6} = h(e_6 e_2^{-1})\}$ (since $T \setminus T_0 = \{e_1\}$), $\{h(e_1)\}$, $h(C^{(1)})$ and $h(C^{(2)})$.

6.3. Strongly singular and singular cases

The next lemma states that if a generalised equation Ω is strongly singular with respect to a periodic structure, then every periodic solution of Ω is in fact a solution of a proper quotient, which can be effectively constructed. In other words, every periodic solution of the generalised equation can be obtained as a composition of a proper epimorphism and a solution of a proper generalised equation.

LEMMA 6.17. *Let $\Omega = \langle \Upsilon, \Re_\Upsilon \rangle$ be a generalised equation without boundary connections, strongly singular with respect to the periodic structure $\langle \mathcal{P}, R \rangle$. Then there exists a finite family of elements $\{g_i\}$, $g_i \in \mathbb{G}_{R(\Omega^*)}$ so that for any solution H of Ω periodic with respect to a period P, such that $\mathcal{P}(H, P) = \langle \mathcal{P}, R \rangle$, one has that $g_i \neq 1$ in $\mathbb{G}_{R(\Omega^*)}$ for some i and $H(g_j) = 1$ for all $g_j \in \{g_i\}$.*

PROOF. Suppose that Ω is strongly singular of type (a) with respect to $\langle \mathcal{P}, R \rangle$. Since Ω is not periodised with respect to $\langle \mathcal{P}, R \rangle$ there exist two cycles $\mathfrak{c}_1, \mathfrak{c}_2 \in C^{(1)} \cup C^{(2)}$ such that $g = [h(\mathfrak{c}_1), h(\mathfrak{c}_2)] \neq 1$ in $\mathbb{G}_{R(\Omega^*)}$. Then, by Lemma 6.8, for every P-periodic solution H of Ω one has $H(g) = 1$. In this case we set the family of elements $\{g_i\}$ to be the finite set of commutators of the form $[h(\mathfrak{c}_1), h(\mathfrak{c}_2)]$, where $\mathfrak{c}_1, \mathfrak{c}_2 \in C^{(1)} \cup C^{(2)}$.

Suppose that Ω is strongly singular of type (b) with respect to $\langle \mathcal{P}, R \rangle$. Notice that by the universal property of the quotient, for every automorphism φ of the coordinate group $\mathbb{G}_{R(\Upsilon^*)}$, the following diagram commutes

$$\begin{array}{ccc} \mathbb{G}_{R(\Upsilon^*)} & \xrightarrow{\varphi} & \mathbb{G}_{R(\Upsilon^*)} \\ \downarrow & & \downarrow \\ \mathbb{G}_{R(\Upsilon^* \cup \mathcal{C})} & \xrightarrow{\tilde{\varphi}} & \mathbb{G}_{R(\Upsilon^* \cup \mathcal{C} \cup \varphi(\mathcal{C}))} \end{array}$$

where $\mathcal{C} = \{[h_i, h_j] \mid \Re_\Upsilon(h_i, h_j)\}$. Note that $\tilde{\varphi}$ is an automorphism of $\mathbb{G}_{R(\Upsilon^* \cup \mathcal{C})} = \mathbb{G}_{R(\Omega^*)}$ if and only if $\varphi(\mathcal{C}) \subset R(\Upsilon^* \cup \mathcal{C})$. Since Ω is strongly singular of type (b) with respect to $\langle \mathcal{P}, R \rangle$, there exists an automorphism φ of $\mathbb{G}_{R(\Upsilon^*)}$ described in statements (2) or (3) of Lemma 6.14 and a commutator $[h_i, h_j] \in \mathcal{C}$ such that $\varphi([h_i, h_j]) \notin R(\Upsilon^* \cup \mathcal{C})$, thus $g = \varphi([h_i, h_j])$ is non-trivial in $\mathbb{G}_{R(\Omega^*)}$. In this case, the finite set of elements of the form $\varphi([h_i, h_j])$, where $\Re_\Upsilon(h_i, h_j)$ and φ is a generator of the group $\mathfrak{A}(\Omega)$, satisfies the required properties.

Since the automorphism $\tilde{\varphi}$ is identical on $\bar{t}$, $h(e)$, where $e \in T_0$, $\bar{u}^{(i)}$, $\bar{z}^{(i)}$, $i = 1, \ldots, m$, and on $h(C^{(1)})$, it follows that $\tilde{\varphi}$ is identical on the set $\{h_k \mid h_k \notin \mathcal{P}\}$. Therefore, from $[h_i, h_j] \in \mathcal{C}$ and $\varphi([h_i, h_j]) \notin R(\Upsilon^* \cup \mathcal{C})$, we get that $h_i \in \mathcal{P}$ or $h_j \in \mathcal{P}$. Without loss of generality, we may assume that $h_i \in \mathcal{P}$.

By part (1) of Lemma 6.14, $\bar{x}$ is a generating set of $\mathbb{G}_{R(\Omega^*)}$. The item h_i can be written in the generators $\bar{x}$ as a word $w(h(\mathrm{Sh}), \bar{u}, \bar{z}, h(C^{(1)}), h(C^{(2)}), h(e_1), \ldots, h(e_m))$. From the above discussion, it follows that $\tilde{\varphi}(h_i) = w(h(\mathrm{Sh}), \bar{u}, \bar{z}, h(C^{(1)}), \tilde{\varphi}(h(C^{(2)})), \tilde{\varphi}(h(e_1)), \ldots, \tilde{\varphi}(h(e_m)))$.

Let H be a solution of Ω periodic with respect to a period P such that $\mathcal{P}(H, P) = \langle \mathcal{P}, R \rangle$. Since

(6.10)
$$\mathrm{alph}(H(h(\mathrm{Sh}))), \mathrm{alph}(H(\bar{u})), \mathrm{alph}(H(\bar{z})), \mathrm{alph}(H(C^{(1)})),$$
$$\mathrm{alph}(H(C^{(2)})), \mathrm{alph}(H(h(e_1))), \ldots, \mathrm{alph}(H(h(e_m))) \subseteq \mathrm{alph}(P),$$

by the definition of the automorphism $\tilde{\varphi}$, we get that

(6.11)
$$\mathrm{alph}(H(h(\mathrm{Sh}))), \mathrm{alph}(H(\bar{u})), \mathrm{alph}(H(\bar{z})), \mathrm{alph}(H(C^{(1)})), \mathrm{alph}(H(\tilde{\varphi}(h(C^{(2)})))),$$
$$\mathrm{alph}(H(\tilde{\varphi}(h(e_1)))), \ldots, \mathrm{alph}(H(\tilde{\varphi}(h(e_m)))) \subseteq \mathrm{alph}(P).$$

Therefore, $\mathrm{alph}(H(\tilde{\varphi}(h_i))) \subseteq \mathrm{alph}(P)$.

Since $h_i \in \mathcal{P}$, we get that $\mathrm{alph}(H_i) = \mathrm{alph}(P)$. As H is a solution of Ω and $\Re_\Upsilon(h_i, h_j)$, so $H_j \leftrightarrows H_i$, i.e. $H_j \leftrightarrows \mathrm{alph}(P)$. Furthermore, from Equation (6.11), it follows that $H(\tilde{\varphi}(h_i)) \leftrightarrows H_j$. Since $\bar{x}$ generates $\mathbb{G}_{R(\Omega^*)}$ and from Equation (6.10), it follows that h_j can be written as a word in $\bar{t}$. Then, since, as mentioned above $\tilde{\varphi}$ is identical on $\bar{t}$, we have that $H(g) = H(\tilde{\varphi}([h_i, h_j])) = H([\tilde{\varphi}(h_i), h_j]) = 1$. □

The next lemma states that if a generalised equation Ω is singular with respect to a periodic structure, then one can construct finitely many proper quotients of the coordinate group $\mathbb{G}_{R(\Omega^*)}$, such that for every periodic solution there exists an $\mathfrak{A}(\Omega)$-automorphic image such that this image is in fact a solution of one of the quotients constructed. In other words, every solution of the generalised equation can be obtained as a composition of an automorphism from $\mathfrak{A}(\Omega)$ and a solution of a proper generalised equation.

LEMMA 6.18. *Let Ω be a generalised equation without boundary connections, singular with respect to the periodic structure $\langle \mathcal{P}, R \rangle$. Then there exists a finite family of cycles $\mathfrak{c}_1, \ldots, \mathfrak{c}_r$ in the graph Γ such that the following conditions hold:*
(1) *$h(\mathfrak{c}_i) \neq 1$, $i = 1, \ldots, r$ in $\mathbb{G}_{R(\Omega^*)}$;*
(2) *for any solution H of Ω periodic with respect to a period P, such that $\mathcal{P}(H, P) = \langle \mathcal{P}, R \rangle$, there exists an $\mathfrak{A}(\Omega)$-automorphic image H^+ of H such that $H^+(\mathfrak{c}_i) = 1$ for some $1 \leq i \leq r$;*
(3) *for any $h_k \notin \mathcal{P}$, $H_k \doteq H_k^+$; and for any $h_k \in \mathcal{P}$ if $H_k \doteq P_1 P^{n_k} P_2$, then $H_k^+ \doteq P_1 P^{n_k^+} P_2$, where $\delta(k) = P_1 P_2$, $n_k, n_k^+ \in \mathbb{Z}$.*

PROOF. Suppose that $|C^{(2)}| \geq 2$. Let H be a solution of a generalised equation Ω periodised with respect to the period P such that $\mathcal{P}(H, P) = \langle \mathcal{P}, R \rangle$. Let $\mathfrak{c}_1, \mathfrak{c}_2 \in C^{(2)}$. By Lemma 6.8, $H(\mathfrak{c}_1) = P^{n_1}$, $H(\mathfrak{c}_2) = P^{n_2}$.

Without loss of generality we may assume that $n_1 \leq n_2$. Write $n_2 = tn_1 + r$, where either $r = 0$ or $0 < |r| < |n_1|$. Applying the automorphism $\varphi : h(\mathfrak{c}_2) \mapsto h(\mathfrak{c}_1)^{-t} h(\mathfrak{c}_2)$, $\varphi \in \mathfrak{A}(\Omega)$, we get $H(\varphi(h(\mathfrak{c}_2))) = P^r$, hence the exponent of periodicity $\exp(H(\varphi(h(\mathfrak{c}_2))))$ is reduced.

Applying the Euclidean algorithm, we get that there exists an automorphism ψ from $\mathfrak{A}(\Omega)$ such that $H(\psi(h(\mathfrak{c}_2))) = 1$ or $H(\psi(h(\mathfrak{c}_1))) = 1$. Set $H^+ = \psi(H)$. Then the set $\{\mathfrak{c}_1, \mathfrak{c}_2\}$ satisfies conditions (1) and (2) of the lemma.

In the notation of Lemma 6.14, the automorphism ψ is identical on all the elements of $\bar{x}$ except for $h(C^{(2)})$, hence, in particular, it is identical on $h(e)$ such that $e \in T$ or $e \notin \mathcal{P}$, and on $h(C^{(1)})$, i.e. $H^+(e) = H(e)$, for any $e \in T$ or $e \notin \mathcal{P}$ and $H^+(\mathfrak{c}^{(1)}) = H(\mathfrak{c}^{(1)})$ for any $\mathfrak{c}^{(1)} \in C^{(1)}$.

If $h_k = h(e)$, $e \notin T$, $h(e) \in \mathcal{P}$, then $h_k = h(\mathfrak{p}_1) h(\mathfrak{c}_e) h(\mathfrak{p}_2)$, where $\mathfrak{p}_1, \mathfrak{p}_2$ are paths in T. Therefore, $H_k^+ = H^+(e) = H^+(\mathfrak{p}_1) H^+(\mathfrak{c}_e) H^+(\mathfrak{p}_2) = H(\mathfrak{p}_1) H^+(\mathfrak{c}_e) H(\mathfrak{p}_2)$. Since $h(\mathfrak{c}_e)$ lies in the subgroup generated by $h(C^{(1)})$ and $h(C^{(2)})$, then $H^+(\mathfrak{c}_e)$ and $H(\mathfrak{c}_e)$ lie in the cyclic group generated by P. This proves statement (3) of the lemma.

Suppose that $C^{(2)} = \{\mathfrak{c}^{(2)}\}$. Since the periodic structure is singular, there exists a cycle $\mathfrak{c}_{e_0}$ such that $h(e_0) \notin \mathcal{P}$, $\mathfrak{c}_{e_0} \in \langle C^{(1)} \rangle$ and $h(\mathfrak{c}_{e_0}) \neq 1$ in $\mathbb{G}_{R(\Omega^*)}$.

We define the set of cycles $\{\mathfrak{c}_1, \ldots, \mathfrak{c}_r\}$ to be

$$\left\{ (\mathfrak{c}_{e_0})^i (\mathfrak{c}^{(2)})^j \mid i \text{ and } j \text{ are not simultaneously zero}, |i|, |j| \leq 2\rho \right\}.$$

We want to show that the elements of the set just defined are non-trivial. In order to do so we show that if $h(\mathfrak{c}_{e_0})^i h(\mathfrak{c}^{(2)})^j = 1$ in $\mathbb{G}_{R(\Omega^*)}$, then $i = j = 0$. Suppose $h(\mathfrak{c}_{e_0})^i h(\mathfrak{c}^{(2)})^j = 1$. Let σ_0 be an automorphism from $\mathfrak{A}(\Omega)$ such that $\sigma_0(h(\mathfrak{c}_{e_0})) = h(\mathfrak{c}_{e_0})$ and $\sigma_0(h(\mathfrak{c}^{(2)})) = h(\mathfrak{c}_{e_0}) h(\mathfrak{c}^{(2)})$. Hence $h(\mathfrak{c}_{e_0})^{i+j} h(\mathfrak{c}^{(2)})^j = 1$ in $\mathbb{G}_{R(\Omega^*)}$, and so $h(\mathfrak{c}_{e_0})^j = 1$. Since $\mathbb{G}$ is torsion-free, this implies that either $h(\mathfrak{c}_{e_0}) = 1$ or $j = 0$. Since $h(\mathfrak{c}_{e_0}) \neq 1$, we have $j = 0$. Then $h(\mathfrak{c}_{e_0})^i = 1$ and, since $\mathbb{G}$ is torsion-free, so $i = 0$.

Let H be a P-periodic solution of the generalised equation Ω such that $\mathcal{P}(H, P) = \langle \mathcal{P}, R \rangle$. By Lemma 6.13, we have $H(\mathfrak{c}_{e_0}) = P^{n_0}$, $|n_0| \leq 2\rho$. Let $H(\mathfrak{c}^{(2)}) = P^n$, $\mathfrak{c}^{(2)} \in C^{(2)}$.

If $n_0 = 0$, we can take $\sigma = 1$, $H^+ = H$ and the set of cycles $\{\mathfrak{c}_{e_0}\}$.

Let $n_0 \neq 0$, $n = tn_0 + n'$, and $|n'| \leq 2\rho$. Let $\sigma = \sigma_0^{-t}$ and define H^+ to be the image of H under σ. Set $\mathfrak{c} = (\mathfrak{c}_{e_0})^{-n'}(\mathfrak{c}^{(2)})^{n_0}$, then $H^+(\mathfrak{c}_{e_0}) = P^{n_0}$, $H^+(\mathfrak{c}^{(2)}) = P^{n'}$ and $H^+(\mathfrak{c}) = 1$. An analogous argument to the one given in the case $|C^{(2)}| \geq 2$ shows that $H_k^+ = H_k$ if $h_k \notin \mathcal{P}$, and $H_k^+ = P_1 P^{n_k^+} P_2$ if $H_k = P_1 P^{n_k} P_2$ and $h_k \in \mathcal{P}$. □

6.4. Example

Let Ω be the generalised equation shown on Figure 12. The associated system of equations is as follows:

$$h_1 h_2 h_3 h_4 = h_4 h_5 h_6 h_7; \ h_1 = h_5; \ h_3 = h_7; \ h_2 = h_8; \ h_6 = h_8; \ h_8 = a.$$

Consider the periodic structure $\langle \mathcal{P}, R \rangle$ defined in Section 6.2. We have shown in the example given in Section 6.2, that this periodic structure is singular. Let $H = (H_1, \ldots, H_8)$ be defined as follows:

$$\begin{array}{llll} H_1 = (bac)^2 b; & H_3 = (cba)^2 c; & H_5 = (bac)^2 b; & H_7 = (cba)^2 c; \\ H_2 = a; & H_4 = (bac)^6; & H_6 = a; & H_8 = a. \end{array}$$

It is easy to see that for $P = bac$ the tuple H is a P-periodic solution of Ω.

In the notation of Lemma 6.14, from the example given in Section 6.2, we have

$$\bar{x} = \left\{ h_8, h_2, h_1 h_6 h_2^{-1} h_1^{-1}, h_6 h_2^{-1}, h(C^{(1)}), h(C^{(2)}) \right\},$$

where $C^{(1)} = \{\mathfrak{c}_{e_3} \mathfrak{c}_{e_7}^{-1}, \mathfrak{c}_{e_5}, \mathfrak{c}_{e_6}\}$, $C^{(2)} = \{\mathfrak{c}_{e_4}, \mathfrak{c}_{e_7}\}$. Notice that, in particular

$$H(C^{(2)}) = \{H(\mathfrak{c}_{e_4}), H(\mathfrak{c}_{e_7})\} = \{P^6, P^5\}.$$

Consider the automorphism $\varphi \in \mathfrak{A}(\Omega)$ induced by the transformation:

$$\mathfrak{c}_{e_4} \mapsto \mathfrak{c}_{e_7}^{-1} \mathfrak{c}_{e_4}; \quad \mathfrak{c}_{e_7} \mapsto (\mathfrak{c}_{e_7}^{-1} \mathfrak{c}_{e_4})^{-5} \mathfrak{c}_{e_7}$$

and identical on the other elements of $\bar{x}$.

Set $H^+ = \varphi(H)$. Then H^+ satisfies the conditions (2) and (3) of Lemma 6.18. Indeed,

$$H^+(\mathfrak{c}_{e_4}) = H(\mathfrak{c}_{e_7}^{-1} \mathfrak{c}_{e_4}) = P \quad H^+(\mathfrak{c}_{e_7}) = H((\mathfrak{c}_{e_7}^{-1} \mathfrak{c}_{e_4})^{-5} \mathfrak{c}_{e_7}) = 1.$$

Furthermore, since φ is the identity on the other elements of $\bar{x}$, we get that

$$H_1^+ = H_1; \ H_2^+ = H_2; \ H_8^+ = H_8;$$

$$H_6^+ H_2^{+^{-1}} = H_6 H_2^{-1} \text{ and so } H_6^+ = H_6;$$

$$H_5^+ H_1^{+^{-1}} = H_5 H_1^{-1} \text{ and so } H_5^+ = H_5;$$

$$H_4^+ = H^+(\mathfrak{c}_{e_4}) = P;$$

$$1 = H^+(\mathfrak{c}_{e_7}) = H_1^+ H_2^+ H_7^+ \text{ and so } H_7^+ = H_2^{-1} H_1^{-1};$$

$$H_3^+ H_7^{+^{-1}} = H_3 H_7^{-1} = 1 \text{ and so } H_3^+ = H_7^+ = H_2^{-1} H_1^{-1}.$$

6.5. Regular case

The aim of this section is to prove that if a generalised equation Ω is regular with respect to a periodic structure, then periodic solutions of Ω minimal with respect to the group of automorphisms $\mathfrak{A}(\Omega)$ have bounded exponent of periodicity. In other words, the length of such a solution is bounded above by a function of Ω and the length $|P|$ of its period P.

LEMMA 6.19. *Let Ω be a generalised equation with no boundary connections, periodised with respect to a connected periodic structure $\langle \mathcal{P}, R \rangle$. Let H be a periodic solution of Ω such that $\mathcal{P}(H, P) = \langle \mathcal{P}, R \rangle$ and minimal with respect to the trivial group of automorphisms. Then, either for all k, $1 \le k \le \rho$ we have*

$$(6.12) \qquad |H_k| \le 2\rho|P|,$$

or there exists a cycle $\mathfrak{c} \in \pi_1(\Gamma, v_\Gamma)$ so that $H(\mathfrak{c}) = P^n$, where $1 \le n \le 2\rho$.

PROOF. Let H be a P-periodic solution of the generalised equation Ω minimal with respect to the trivial group of automorphisms.

Suppose that there exists a variable $h_k \in \mathcal{P}$ such that $|H_k| > 2\rho|P|$. We then prove that there exists a cycle $\mathfrak{c} \in \pi_1(\Gamma, v_\Gamma)$ so that $H(\mathfrak{c}) = P^n$, where $1 \le n \le 2\rho$.

Construct a chain

$$(6.13) \qquad (\Omega, H) = (\Omega_{v_0}, H^{(0)}) \to (\Omega_{v_1}, H^{(1)}) \to \cdots \to (\Omega_{v_t}, H^{(t)}),$$

in which for all i, $\Omega_{v_{i+1}}$ is obtained from Ω_{v_i} using ET 5: by μ-tying a free boundary that intersects a base $\mu \in \mathcal{P}$. Chain (6.13) is constructed once all the boundaries intersecting any of the bases μ from $\mathcal{P}$ are μ-tied. This chain is finite, since, by the definition, boundaries that are introduced when applying ET 5 are not free.

By construction, the generalised equations Ω_{v_i}'s in (6.13) have boundary connections. Our definition of periodic structure is given for generalised equations without boundary connections, see Definition 6.3. For this reason we define $\Omega_{v_i'}$ to be the generalised equation obtained from Ω_{v_i} by omitting all boundary connections (here we do not apply D 3), but leaving all the boundaries that may have been introduced when ET 5 was applied. It is obvious that the solution $H^{(i)}$ of Ω_{v_i} is also a solution of the generalised equation $\Omega_{v_i'}$ and is periodic with respect to the period P. Denote by $\langle \mathcal{P}_i, R_i \rangle$ the periodic structure $\mathcal{P}(H^{(i)}, P)$ on the generalised equation $\Omega_{v_i'}$ restricted to the closed sections from $\mathcal{P}$, and by $\Gamma^{(i)}$ the corresponding graph.

If (p, μ, q), $\mu \in \mathcal{P}$, is a boundary connection of the generalised equation Ω_{v_i}, $i = 1, \ldots, t$, then $\delta(p) = \delta(q)$. Therefore, all the graphs $\Gamma^{(0)}, \Gamma^{(1)}, \ldots, \Gamma^{(t)}$ have the same set of vertices, whose cardinality does not exceed ρ. By Lemma 5.8, the solution $H^{(t)}$ of the generalised equation Ω_{v_t} is also minimal with respect to the trivial group of automorphisms.

Suppose that for some variable h_l (of Ω_{v_t}) belonging to a section from $\mathcal{P}$ the inequality $|H_l^{(t)}| > 2|P|$ holds.

Let $\mathcal{H} = \left\{ h_i \in \sigma \mid \sigma \in \mathcal{P} \text{ and } H_i^{(t)} \doteq H_l^{(t) \pm 1} \right\}$ be a set of items of Ω_{v_t}. Consider the partially commutative $\mathbb{G}$-group $\mathbb{G}(\mathcal{A} \cup \{u\})$, where u is a new letter and $[u, a_k] = 1$, $a_k \in \mathcal{A}$ for all $a_k \in \mathbb{A}(H_j)$ (see Section 2.5 for definition), where $\Re_\Upsilon(h_i, h_j)$ for some $h_i \in \mathcal{H}$.

6.5. REGULAR CASE

In the solution $H^{(t)}$, replace all the components $H_i^{(t)}$ such that $H_i^{(t)} \doteq H_l^{(t)}$ or $H_i^{(t)} \doteq H_l^{(t)-1}$ by the letter u or u^{-1}, correspondingly (see Definition 5.6). Denote the resulting $\rho_{\Omega_{v_t}}$-tuple of words by $H^{(t)'}$.

We show that $H^{(t)'}$ is in fact a solution of Ω_{v_t} (considered as a generalised equation with coefficients from $\mathcal{A} \cup \{u\}$, see Remark 5.5). Obviously, by definition, every component of $H^{(t)'}$ is non-empty and geodesic. Since in the generalised equation Ω_{v_t} all the boundaries from $\mathcal{P}$ are μ-tied, $\mu \in \mathcal{P}$, there is a one-to-one correspondence between the items that belong to μ and the items that belong to $\Delta(\mu)$. Therefore, $H^{(t)'}$ satisfies all basic equations $h(\mu) = h(\Delta(\mu))$, $\mu \in \mathcal{P}$ and all the boundary equations of the generalised equation Ω_{v_t}.

If, on the other hand, $\mu \notin \mathcal{P}$, then for every item $h_k \in \mu$ of the generalised equation Ω lying on a closed section from $\mathcal{P}$, we have $h_k \notin \mathcal{P}$ and, consequently, $|H_k| \leq 2|P|$. In particular, this inequality holds for every item $h_k \in \mu$, $\mu \notin \mathcal{P}$ of the generalised equation Ω_{v_t}. Therefore, such items have not been replaced in the solution $H^{(t)}$, thus $H^{(t)'}$ satisfies all basic equations $h(\mu) = h(\Delta(\mu))$, $\mu \notin \mathcal{P}$. From the construction of $\mathbb{G}(\mathcal{A} \cup \{u\})$ it follows that $H^{(t)'}$ satisfies the constraints from $\Re_{\Upsilon_{v_t}}$.

Let $\pi : \mathbb{G}(\mathcal{A} \cup \{u\}) \to \mathbb{G}(\mathcal{A} \cup \{u\})$ be a map that sends u to $H_l^{(t)}$ and fixes $\mathbb{G}$. Since $H^{(t)}$ is a solution of Ω_{v_t}, one has $H_l^{(t)} \leftrightarrows H_j^{(t)}$, where $\Re_{\Upsilon_{v_t}}(h_i, h_j)$, $h_i \in \mathcal{H}$. Therefore, $[H_l^{(t)}, a_k] = 1$, where $a_k \in \text{alph}(H_j^{(t)})$ and $\Re_{\Upsilon_{v_t}}(h_i, h_j)$, $h_i \in \mathcal{H}$. These are the relations that, by construction, are satisfied by u, hence π is a $\mathbb{G}$-endomorphism. This contradicts the minimality of the solution $H^{(t)}$. Indeed, one has $\pi_{H^{(t)}} = \pi_{H^{(t)'}} \pi$. By construction, it is easy to check that conditions (2) and (3) of Definition 5.1 hold. Therefore, $H^{(t)'} <_{\{1\}} H^{(t)}$. Obviously, $|H_i^{(t)'}| = |H_i^{(t)}|$ for all $i \neq l$ and $1 = |u| = |H_l^{(t)'}| < |H_l^{(t)}|$. We thereby have shown that $|H_l^{(t)}| \leq 2|P|$, if h_l belongs to a closed section from $\mathcal{P}$.

In the construction of chain (6.13) we introduced new boundaries, so every item from Ω_{v_0} can be expressed as a product of items from $\Omega_{v'_t}$. Consequently, since $|H_l^{(t)}| \leq 2|P|$ for every l, if the component H_k of the solution H of the generalised equation Ω does not satisfy inequality (6.12), then h_k is a product of at least $\rho + 1$ distinct items of $\Omega_{v'_t}$, $h_k = h_s^{(t)} \cdots h_{s+\varrho}^{(t)}$, where $\varrho \geq \rho + 1$.

Since the graph $\Gamma^{(t)}$ contains at most ρ vertices, there exist boundaries $l, l' \in \{s, \ldots, s + \rho + 1\}$, $l < l'$ so that $\delta(l) = \delta(l')$. The word $h[l, l']$ is a label of a cycle $\mathfrak{c}_t$ of the graph $\Gamma^{(t)}$ for which $0 < |H(\mathfrak{c}_t)| \leq 2\rho|P|$.

Recall that by $\pi(v_i, v_j)$ we denote the homomorphism $\mathbb{G}_{R(\Omega_{v_i}*)} \to \mathbb{G}_{R(\Omega_{v_j}*)}$ induced by the elementary transformations. It remains to prove the existence of a cycle $\mathfrak{c}_0$ of the graph Γ for which $\pi(v_\Gamma, v_t)(h(\mathfrak{c}_0)) = h(\mathfrak{c}_t)$. To do this, it suffices to show that for every path $\mathfrak{p}_{i+1} : v \to v'$ in the graph $\Gamma^{(i+1)}$ there exists a path $\mathfrak{p}_i : v \to v'$ in the graph $\Gamma^{(i)}$ such that $\pi(v_i, v_{i+1})(h(\mathfrak{p}_i)) = h(\mathfrak{p}_{i+1})$. In turn, it suffices to verify the latter statement in the case when $\mathfrak{p}_{i+1}$ is an edge e.

The generalised equation $\Omega_{v_{i+1}}$ is obtained from Ω_{v_i} by μ-tying a boundary p. Below we use the notation from the definition of the elementary transformation ET 5.

(1) Either we introduce a boundary connection (p, μ, q). In this case every variable $h(e)$ of the generalised equation $\Omega_{v_{i+1}}$ is also a variable of the generalised equation Ω_{v_i} and the statement is obvious;
(2) Or we introduce a new boundary q' between the boundaries q and $q+1$, and a boundary connection (p, μ, q'). Using the boundary equations we get

(6.14)
$$\pi(v_i, v_{i+1})^{-1}(h_q) = h[\alpha(\Delta(\mu)), q]^{-1} h[\alpha(\Delta(\mu)), q'] = h[\alpha(\Delta(\mu)), q]^{-1} h[\alpha(\mu), p],$$
$$\pi(v_i, v_{i+1})^{-1}(h_{q'}) = h[\alpha(\Delta(\mu)), q']^{-1} h[\alpha(\Delta(\mu)), q+1] =$$
$$h[\alpha(\mu), p]^{-1} h[\alpha(\Delta(\mu)), q+1].$$

Moreover, $\pi(v_i, v_{i+1})^{-1}(h(e)) = h(e)$ for any other variable. Notice that since $(\alpha(\mu)) = (\alpha(\Delta(\mu)))$, the right-hand sides of Equations (6.14) are labels of paths in $\Gamma^{(i)}$.

Thus, we have deduced that there exists a cycle $\mathfrak{c} \in \Gamma$ so that $1 \leq \exp(H(\mathfrak{c})) \leq 2\rho$. $\square$

LEMMA 6.20. *Let Ω be a generalised equation with no boundary connections. Suppose that Ω is regular with respect to a periodic structure $\langle \mathcal{P}, R \rangle$. Then there exists a computable function $f_0(\Omega, \mathcal{P}, R)$ such that, for every $\mathcal{P}$-periodic solution H of Ω such that $\mathcal{P}(H, P) = \langle \mathcal{P}, R \rangle$ and such that H is minimal with respect to $\mathfrak{A}(\Omega)$, the following inequality holds*

$$|H_k| \leq f_0(\Omega, \mathcal{P}, R) \cdot |P| \text{ for every } k.$$

PROOF. Let H be a $\mathcal{P}$-periodic minimal solution with respect to the group of automorphisms $\mathfrak{A}(\Omega)$. Notice that H is also minimal with respect to the trivial group of automorphisms.

We first prove that for any regular periodic structure and any $\mathcal{P}$-periodic solution H of Ω minimal with respect to the group of automorphisms $\mathfrak{A}(\Omega)$, the exponent of periodicity of the label of every simple cycle $\mathfrak{c}_e$, $e \notin T$ is bounded by a certain computable function $g_1(\Omega, \mathcal{P}, R)$.

If for every i we have $|H_i| \leq 2\rho|P|$, the statement follows. We further assume that there exists i such that $|H_i| > 2\rho|P|$.

Since the periodic structure is regular, either $|C^{(2)}| = 1$ and for all $e \notin T$ so that $e \in \text{Sh}$ we have $H(\mathfrak{c}_e) = 1$, or $|C^{(2)}| = 0$.

Assume first that, $C^{(2)} = \{\mathfrak{c}^{(2)}\}$. Since the periodic structure $\langle \mathcal{P}, R \rangle$ is regular, it follows that $H(\mathfrak{c}_e) = 1$ for all $e \notin T$, $h(e) \notin \mathcal{P}$. Since H is a solution of Ω, it follows that $H(b_\mu) = 1$ and hence $\exp(H(b_\mu)) = 0$, $\mu \in \mathcal{P}$. Therefore, $H(\mathfrak{c}) = 1$ and hence $\exp(H(\mathfrak{c})) = 0$ for all $\mathfrak{c}$ such that $\tilde{\mathfrak{c}} \in \widetilde{B}$. By (6.4), we have that $[\widetilde{Z}_1 : \widetilde{B}] < \infty$, hence we get that $\exp(H(\mathfrak{c})) = 0$ for all $\mathfrak{c}$ such that $\tilde{\mathfrak{c}} \in \widetilde{Z}_1$. Consequently, since the only non-trivial cycle at v_Γ is $\mathfrak{c}^{(2)}$, by Lemma 6.19, we get that $|\exp(H(\mathfrak{c}^{(2)}))| \leq 2\rho$. Using factorisation (6.4), one can effectively express $\tilde{\mathfrak{c}}_e$, $e \notin T$ in terms of the elements of the basis:

$$\tilde{\mathfrak{c}}_e = n_e \tilde{\mathfrak{c}}^{(2)} + \tilde{z}_e^{(1)}, \ \tilde{z}_e^{(1)} \in \widetilde{Z}_1.$$

Note that the number n_e depends only on the generalised equation Ω (and does not depend on the solution H). From the above discussion and Remark 6.10, it follows

6.5. REGULAR CASE

that $|\exp(H(\mathfrak{c}_e))| = |n_e \exp(H(\mathfrak{c}^{(2)}))| \leq 2\rho n_e$, and we finally obtain

(6.15) $$|\exp(H(\mathfrak{c}_e))| \leq g_1(\Omega, \mathcal{P}, R),$$

where g_1 is a certain computable function.

Suppose next that $|C^{(2)}| = 0$, i.e. $\widetilde{Z} = \widetilde{Z}_1$. As we have already seen in the proof of Lemma 6.13, $|H(\mathfrak{c}_e)| \leq 2\rho|P|$, where $e \notin T$, $h(e) \notin \mathcal{P}$. Hence, $|\exp(H(\mathfrak{c}_e))| \leq 2\rho$ for all e such that $h(e) \notin \mathcal{P}$. Since $(\widetilde{Z}_1 : \widetilde{B}) < \infty$, for every $e_0 \notin T$ one can effectively construct the following equality

$$n_{e_0}\tilde{\mathfrak{c}}_{e_0} = \sum_{h(e) \notin \mathcal{P}} n_e \tilde{\mathfrak{c}}_e + \sum_{\mu \in \mathcal{P}} n_\mu \tilde{b}_\mu.$$

Hence,

$$|\exp(H(\mathfrak{c}_{e_0}))| \leq |n_{e_0} \cdot \exp(H(\mathfrak{c}_{e_0}))| \leq \sum_{h(e) \notin \mathcal{P}} |n_e \exp(H(\mathfrak{c}_e))| \leq 2\rho \cdot \sum_{h(e) \notin \mathcal{P}} |n_e|.$$

Thus, $|\exp(H(\mathfrak{c}_{e_0}))| \leq g_2(\Omega, \mathcal{P}, R)$.

We now address the statement of the lemma. The way we proceed is as follows. For a P-periodic solution H minimal with respect to the group of automorphisms $\mathfrak{A}(\Omega)$, we show that the vector that consists of exponents of periodicity of each of the components H_k of H is bounded by a minimal solution of a linear system of equations whose coefficients depend only on the generalised equation. Since by Lemma 1.1 from [**Mak77**], the components of a minimal solution of a linear system of equations are bounded above by a recursive function of the coefficients of the system, we then get a recursive bound on the exponents of periodicity of the components of the solution H.

Let $\delta(k) = P_1^{(k)} P_2^{(k)}$. Denote by $t(\mathfrak{c}, h_k)$ the algebraic sum of occurrences of the edge with the label h_k in the cycle $\mathfrak{c}$, (i.e. edges with different orientation contribute with different signs). For every item h_k that belongs to a closed section from $\mathcal{P}$ one can write

$$H_k = P_2^{(k)} P^{n_k} P_1^{(k+1)}.$$

Note that in the case that $h_k \in \mathcal{P}$, the above equality is graphical. However, in the case that $h_k \notin \mathcal{P}$ and H_k is a subword of $P^{\pm 1}$ there is cancellation. Direct calculations show that

(6.16) $$H(\mathfrak{c}) = P^{\left(\sum_k t(\mathfrak{c}, h_k)(n_k + 1)\right) - 1}.$$

By Lemma 6.19, $e_0 \notin T$ can be chosen in such a way that $\exp(H(\mathfrak{c}_{e_0})) \neq 0$. Let $n_k = |\exp(\tilde{\mathfrak{c}}_{e_0})| m_k + r_k$, where $0 < r_k \leq |\exp(\tilde{\mathfrak{c}}_{e_0})|$. Equation (6.16) implies that the vector $\{m_k \mid h_k \in \mathcal{P}\}$ is a solution of the following system of Diophantine equations in variables $\{z_k \mid h_k \in \mathcal{P}\}$:

(6.17)
$$\sum_{h_k \in \mathcal{P}} t(\mathfrak{c}_e, h_k)(|\exp(H(\mathfrak{c}_{e_0}))| z_k + r_k + 1) + \sum_{h_k \notin \mathcal{P}} t(\mathfrak{c}_e, h_k)(n_k + 1) - 1 = \exp(H(\mathfrak{c}_e)),$$

where $e \notin T$. Note that the number of variables of the system (6.17) is bounded. Furthermore, as we have proven above, free terms $\exp(H(\mathfrak{c}_e))$ of this system are also bounded above, and so are the coefficients $|n_k| \leq 2$ for $h_k \notin \mathcal{P}$.

A solution $\{m_k\}$ of a system of linear Diophantine equations is called *minimal*, see [**Mak77**], if $m_k \geq 0$ and there is no other solution $\{m_k^+\}$ such that $0 \leq m_k^+ \leq m_k$

for all k, and at least one of the inequalities $m_k^+ \leq m_k$ is strict. Let us verify that the solution $\{m_k \mid h_k \in \mathcal{P}\}$ of system (6.17) is minimal.

Indeed, let $\{m_k^+\}$ be another solution of system (6.17) such that $0 \leq m_k^+ \leq m_k$ for all k, and at least for one k the inequality is strict. Let $n_k^+ = |\exp(H(\mathfrak{c}_{e_0}))|m_k^+ + r_k$. Define a vector H^+ as follows: $H_k^+ = H_k$ if $h_k \notin \mathcal{P}$, and $H_k^+ = P_2^{(k)} P^{n_k^+} P_1^{(k+1)}$ if $h_k \in \mathcal{P}$.

We now show that H^+ is a solution of the generalised equation which can be obtained from H by an automorphism from $\mathfrak{A}(\Omega)$.

The vector H^+ satisfies all the coefficient equations and all the basic equations of Ω. Indeed, since $\{m_k^+\}$ is a solution of system (6.17), $H^+(\mathfrak{c}_e) = P^{\exp(H(\mathfrak{c}_e))} = H(\mathfrak{c}_e)$. Therefore, for every cycle $\mathfrak{c}$ we have $H^+(\mathfrak{c}) = H(\mathfrak{c})$ and, in particular, $H^+(b_\mu) = H(b_\mu) = 1$. Thus the vector H^+ is a solution of the system Ω^*.

By construction every component of the solution H^+ is non-empty and the words $H^+(\mu)$, $H^+(\Delta(\mu))$ are geodesic. On the other hand, for every μ we have

$$H^+(\mu) H^+(\Delta(\mu))^{-1} = 1.$$

Since if $\mu \in \mathcal{P}$, then the words $H^+(\mu)$ and $H^+(\Delta(\mu))$ are $\mathcal{P}$-periodic, it follows that

$$H^+(\mu) \doteq H^+(\Delta(\mu)).$$

From the definition of H^+ it is obvious that H^+ satisfies the commutation constraints induced by $\mathfrak{R}_\Upsilon$. Thus, H^+ is a solution of the generalised equation Ω.

Denote by δ_{ie_0} an element from the group of automorphisms $\mathfrak{A}(\Omega)$ defined in the following way. For every $e_i \in T \setminus T_0$, $i = 1, \ldots, m$ set

$$\delta_{ie_0} : h(e_j) \mapsto \begin{cases} h(\mathfrak{p}(v_\Gamma, v_i)^{-1} \mathfrak{c}_{e_0} \mathfrak{p}(v_\Gamma, v_i)) h(e_i), & \text{for } j = i; \\ h(e_j), & \text{for } j \neq i. \end{cases}$$

Therefore, if $\pi_{H'} = \delta_{ie_0} \pi_H$ and $h(e_i) = h_k \in \mathcal{P}$, then

$$H_k' = P_2^{(k)} P^{n_k + \exp(H(\mathfrak{c}_{e_0}))} P_1^{(k+1)},$$

and all the other components of H' are the same as in H, $H_l' = H_l$, $l \neq k$. Denote by $\delta = \prod_{i=1}^{m} \delta_{ie_0}^{\Delta_i}$, where $h(e_i) = h_{k_i}$, $\Delta_i = (m_{k_i}^+ - m_{k_i}) \cdot \text{sign}(\exp(H(\mathfrak{c}_{e_0})))$. Let us verify the equality

(6.18) $$\pi_{H^+} = \pi_H \delta.$$

Let $\pi_{H^{(1)}} = \delta \pi_H$. Then, by construction, $H_k^{(1)} = P_2^{(k)} P^{m_k^+} P_1^{(k+1)} = H_k^+$ for all k such that h_k is the label of an edge from $T \setminus T_0$. If the edge labelled by h_k belongs to T_0, or h_k does not belong to a closed section from $\mathcal{P}$, then $h_k \notin \mathcal{P}$ and $H_k^{(1)} = H_k = H_k^+$. Finally, for every $e \notin T$ we have $H^{(1)}(\mathfrak{c}_e) = H(\mathfrak{c}_e) = H^+(\mathfrak{c}_e)$. As $\mathfrak{c}_e = \mathfrak{p}_1 e \mathfrak{p}_2$, where $\mathfrak{p}_1, \mathfrak{p}_2$ are paths in the tree T, and for every item h_k which labels an edge from T, the equality $H_k^{(1)} = H_k^+$ has already been established, so it follows that $H^{(1)}(e) = H^+(e)$. This proves (6.18).

Since H^+ is a solution of Ω, since by construction H^+ satisfies conditions (2) and (3) from Definition 5.1 and by (6.18), it follows that $H^+ <_{\mathfrak{A}(\Omega)} H$. We arrive to a contradiction with the minimality of the solution H. Consequently, the solution $\{m_k \mid h_k \in \mathcal{P}\}$ of system (6.17) is minimal.

Lemma 1.1 from [**Mak77**] states that the components of the minimal solution $\{m_k \mid h_k \in \mathcal{P}\}$ are bounded by a recursive function of the coefficients, the number of variables and the number of equations of the system. Since, as shown above, all of these parameters of system (6.17) are bounded above by a computable function, the lemma follows. □

CHAPTER 7

The finite tree $T_0(\Omega)$ and minimal solutions

In Chapter 4 we constructed an infinite tree $T(\Omega)$. Though for every solution H the path in $T(\Omega)$

$$(\Omega, H) = (\Omega_{v_0}, H^{(0)}) \to (\Omega_{v_1}, H^{(1)}) \to \cdots \to (\Omega_{v_t}, H^{(t)})$$

defined by the solution H is finite, in general, there is no global bound for the length of these paths.

Informally, the aim of this chapter is to prove that the set of solutions of the generalised equation Ω can be parametrised by a finite subtree T_0 (to be constructed below) of the tree T, automorphisms from a recursive group of automorphisms of the coordinate group $\mathbb{G}_{R(\Omega^*)}$, and solutions of the generalised equations associated to leaves of T_0. In other words, we prove that there exists a global bound M, such that for any solution H of Ω one can effectively construct a path $\mathfrak{p}(H) = (\Omega_{v_0}, H^{[0]}) \to (\Omega_{v_1}, H^{[1]}) \to \ldots$ in $T(\Omega)$ of length bounded above by M and such that $H^{[i]} <_{\text{Aut}(\Omega)} H$ for all i, where $\text{Aut}(\Omega)$ is the group of automorphisms of $\mathbb{G}_{R(\Omega^*)}$ defined in Section 7.2, see Definition 7.16 (note the abuse of notation: $H^{[i]}$ and H are solutions of different generalised equations; in fact this relation is between two solutions of Ω: a solution induced by $H^{[i]}$ and H; see below for a formal definition).

We summarise the results of this chapter in the proposition below.

PROPOSITION 7.1. *For a (constrained) generalised equation $\Omega = \Omega_{v_0}$, one can effectively construct a finite oriented rooted at v_0 tree T_0, $T_0 = T_0(\Omega_{v_0})$, such that:*
(1) *The tree T_0 is a subtree of the tree $T(\Omega)$.*
(2) *To the root v_0 of T_0 we assign a recursive group of automorphisms $\text{Aut}(\Omega)$ (see Definition 7.16).*
(3) *For any solution H of a generalised equation Ω there exists a leaf w of the tree $T_0(\Omega)$, $\text{tp}(w) = 1, 2$, and a solution $H^{[w]}$ of the generalised equation Ω_w such that*
  - $H^{[w]} <_{\text{Aut}(\Omega)} H$;
  - *if $\text{tp}(w) = 2$ and the generalised equation Ω_w contains non-constant non-active sections, then there exists a period P such that $H^{[w]}$ is periodic with respect to the period P and the generalised equation Ω_w is either singular of strongly singular with respect to the periodic structure $\mathcal{P}(H^{[w]}, P)$.*

This chapter is organised in three sections. The aim of Section 7.1 is to define the recursive group of automorphisms $\mathfrak{V}(\Omega_v)$ of the coordinate group of the generalised equation Ω_v associated to v, $v \in T(\Omega)$.

In Section 7.2, we define a finite subtree $T_0(\Omega)$ of the tree $T(\Omega)$. In order to define $T_0(\Omega)$ we introduce the notions of prohibited paths of type 7-10, 12 and 15.

Using Lemma 4.19, we prove that any infinite branch of the tree $T(\Omega)$ contains a prohibited path. The tree $T_0(\Omega)$ is defined to be a subtree of $T(\Omega)$ that does not contain prohibited paths and, by construction, is finite. We then define a recursive group of automorphisms $\mathrm{Aut}(\Omega)$ that we assign to the root vertex of the tree $T_0(\Omega)$. The group $\mathrm{Aut}(\Omega)$ is generated by conjugates of the groups $\mathfrak{V}(\Omega_v)$, $v \in T_0(\Omega)$, $\mathrm{tp}(v) \neq 1$.

Finally, in Section 7.3 for any solution H of Ω we construct the path $\mathfrak{p}(H)$ from the root v_0 to w, prove that w is a leaf of the tree $T_0(\Omega)$ and show that the leaf w satisfies the properties required in Proposition 7.1.

7.1. Automorphisms

One of the features of our results in this paper is that all the constructions we give are effective. Since, we do not have control over the automorphism groups of $\mathbb{G}$-residually $\mathbb{G}$ groups, we are led to narrow the group of automorphisms in such a way that on the one hand they are effectively described and, on the other hand, Proposition 7.1 stated above still holds.

Informally, we will see that the automorphisms that we consider (besides the ones already defined in Chapter 6 for periodic structures) are, in a sense, induced by the canonical epimorphisms in the infinite branches of the tree T. Therefore, these automorphisms are compositions of epimorphisms induced by elementary and derived transformations. This will allow us to prove that the automorphisms are completely induced: come from word mappings that, in turn, induce automorphisms of residually free groups considered by Razborov in [**Raz87**]. Since the group of automorphisms considered by Razborov are tame and finitely generated, using the decidability of the universal theory of $\mathbb{G}$, we will get that our groups of automorphisms are recursive.

In his work on systems of equations over a free group [**Raz85**], [**Raz87**], to prove that the groups of automorphisms are finitely generated, Razborov uses a result of McCool, see [**McC75**]. When this paper was already written M. Day published two preprints, [**Day08a**] and [**Day08b**] on automorphisms of partially commutative groups. The authors believe that using ideas of A. Razborov and techniques developed by M. Day, one can prove that the automorphisms groups of the coordinate groups we consider are, in fact, finitely generated.

DEFINITION 7.2. Let $\Omega = \langle \Upsilon, \Re_\Upsilon \rangle$ be a generalised equation. An automorphism θ of the coordinate group $\mathbb{G}_{R(\Omega^*)}$ is called *completely induced* if there exist an F-homomorphism $\tilde{\theta}$ from $F[h]$ to $F[h]$, where F is the free group on $\mathcal{A}$ and an F-automorphism θ' of the coordinate group $F_{R(\Upsilon^*)}$, such that the following diagram commutes:

$$\begin{array}{ccccc}
\mathbb{G}_{R(\Omega^*)} & \longleftarrow & F[h] & \longrightarrow & F_{R(\Upsilon^*)} \\
\theta \downarrow & & \tilde{\theta} \downarrow & & \downarrow \theta' \\
\mathbb{G}_{R(\Omega^*)} & \longleftarrow & F[h] & \longrightarrow & F_{R(\Upsilon^*)}
\end{array}$$

In this case, the automorphisms θ and θ' are called *dual*.

To every vertex v of the tree $T(\Omega)$, $\Omega = \langle \Upsilon, \Re_\Upsilon \rangle$ we assign a recursive group of automorphisms $\mathfrak{V}(\Omega_v)$ of $\mathbb{G}_{R(\Omega_v^*)}$.

7.1. AUTOMORPHISMS

If $\mathrm{tp}(v) = 2$, set $\mathfrak{V}(\Omega_v)$ to be the group generated by all the groups of automorphisms $\mathfrak{A}(\Omega_v)$ corresponding to *regular* periodic structures on Ω_v, see Definition 6.16.

REMARK 7.3. The definition of the automorphism groups $\mathfrak{V}(\Omega_v)$ in the case when $7 \leq \mathrm{tp}(v) \leq 10$ (or, $\mathrm{tp}(v) = 12$) given below may seem unnatural. In fact, for the purposes of this paper, the automorphism group $\mathfrak{V}(\Omega_v)$ could be defined as the group of automorphisms of $\mathbb{G}_{R(\Omega_v^*)}$ that are induced by *F-homomorphisms* φ of the free group $F[h]$ such that $\varphi(h_i) = h_i$ for all h_i that belong to the kernel of Ω_v (or, such that $\varphi(h_i) = h_i$ for all h_i that belong to the non-quadratic part of Ω_v, see below for definition). If defined in this way, the groups of automorphisms $\mathfrak{V}(\Omega_v)$ are recursive and all the proofs in the paper work in the same way.

However, we chose to define the groups $\mathfrak{V}(\Omega_v)$ as below because the groups of automorphisms defined by Razborov are well understood (in the case that the corresponding coordinate group is fully residually free they correspond to the canonical group of automorphisms associated to its JSJ decomposition, see [**KhM98b**] and [**KhM05b**]) and we hope that establishing relations between $\mathfrak{V}(\Omega_v)$ and the groups defined by Razborov will help understanding the structure of $\mathfrak{V}(\Omega_v)$ and, consequently, of the splittings of fully residually $\mathbb{G}$ groups.

For the time being, the only consequences of our definition are that the group of automorphisms $\mathfrak{V}(\Omega_v)$ is "induced" by a subgroup of the automorphism groups defined by Razborov and that automorphisms from $\mathfrak{V}(\Omega_v)$ are tame. Recall that an automorphism of a group is *tame* if it is induced by an automorphism of the free group.

DEFINITION 7.4. A tame group of automorphisms of a group $G = \langle X \mid R \rangle$ is called *recursive* if it is recursive when viewed as a subset of $\mathrm{Aut}(F(X))$.

Let $7 \leq \mathrm{tp}(v) \leq 10$. An automorphism φ of the coordinate group $\mathbb{G}_{R(\Omega_v^*)}$ is called *invariant with respect to the kernel* if it is completely induced (see Definition 7.2) and its dual is invariant with respect to the kernel in the sense of Razborov, [**Raz87**]. We now recall the definition of an automorphism invariant with respect to the kernel in the sense of Razborov.

Let Υ_v be a non-constrained generalised equation over the free monoid $\mathbb{F}$, and $F_{R(\Upsilon_v^*)}$ be its coordinate group over F, where F is the free group with basis $\mathcal{A}$. Let $\widehat{\Upsilon}_v$ be obtained from $D\,3(\Upsilon_v)$ by removing all coefficient equations and all bases from the kernel of Υ_v. Let π be the natural homomorphism from $F_{R(\widehat{\Upsilon}_v^*)}$ to $F_{R(\Upsilon_v^*)}$.

An automorphism φ of the coordinate group $F_{R(\Upsilon_v^*)}$ is called *invariant with respect to the kernel*, see [**Raz87**], if it is induced by an automorphism φ' of the coordinate group $F_{R(\widehat{\Upsilon}_v^*)}$ identical on the kernel of Υ_v, i.e. there exists an F-automorphism φ' of the coordinate group $F_{R(\widehat{\Upsilon}_v^*)}$ so that $\varphi'(h_i) = h_i$ for every variable $h_i \in \mathrm{Ker}(\Upsilon_v)$ and the following diagram commutes

$$\begin{array}{ccc} F_{R(\widehat{\Upsilon}_v^*)} & \xrightarrow{\pi} & F_{R(\Upsilon_v^*)} \\ \varphi' \downarrow & & \varphi \downarrow \\ F_{R(\widehat{\Upsilon}_v^*)} & \xrightarrow{\pi} & F_{R(\Upsilon_v^*)} \end{array}$$

LEMMA 7.5 (Lemma 2.5, [**Raz87**]). *There exists a finitely presented subgroup K of the group of automorphisms of a free group such that every automorphism from K*

induces an automorphism of $F^{[h]}/\mathrm{ncl}\langle \Upsilon_v \rangle$, which, in turn, induces an automorphism of the coordinate group $F_{R(\Upsilon_v^*)}$. The finitely generated group of all automorphisms K' of $F_{R(\Upsilon_v^*)}$ induced by automorphisms from K contains all the automorphisms invariant with respect to the kernel of Υ_v.

We use the notation of Lemma 7.5. Every automorphism of $\mathbb{G}_{R(\Omega_v^*)}$ invariant with respect to the kernel is completely induced and, by definition, its dual belongs to K'. For every automorphism φ from K, there exists an algorithm to decide whether or not φ induces an automorphism of $\mathbb{G}_{R(\Omega_v^*)}$. Indeed, in order to do so it suffices to check if the following finite family of universal formulas (quasi-identities) holds in $\mathbb{G}$:

$$\left\{\begin{array}{ll} \forall H\ [\varphi(\Upsilon^*(H))=1 \wedge \{\varphi([H_i,H_j])=1 \mid \Re_\Upsilon(h_i,h_j)\}] & \to\ u(H)=1; \\ \forall H\ [\varphi(\Upsilon^*(H))=1 \wedge \{\varphi([H_i,H_j])=1 \mid \Re_\Upsilon(h_i,h_j)\}] & \to\ [H_i,H_j]=1; \\ \forall H\ [\Upsilon^*(H)=1 \wedge \{[H_i,H_j]=1 \mid \Re_\Upsilon(h_i,h_j)\}] & \to\ \varphi(u(H))=1; \\ \forall H\ [\Upsilon^*(H)=1 \wedge \{[H_i,H_j]=1 \mid \Re_\Upsilon(h_i,h_j)\}] & \to\ \varphi([H_i,H_j])=1. \end{array}\right\}$$

for every equation u of the system Υ^* and every pair (H_i, H_j) such that $\Re_\Upsilon(h_i, h_j)$. This can be done effectively, since the universal theory of the group $\mathbb{G}$ is decidable, see [**DL04**]. We therefore get the following lemma.

LEMMA 7.6. *The group of all automorphisms of $\mathbb{G}_{R(\Omega_v^*)}$ invariant with respect to the kernel of Ω_v is tame and recursive.*

In this case, we define $\mathfrak{V}(\Omega_v)$ to be the group of automorphisms of $\mathbb{G}_{R(\Omega_v^*)}$ invariant with respect to the kernel.

Let $\mathrm{tp}(v) = 15$. Apply derived transformation D 3 and consider the generalised equation $D\,3(\Omega_v) = \widetilde{\Omega}_v$. Notice that, since every boundary in the active part of $\widetilde{\Omega}_v$ that touches a base and intersects another base μ is μ-tied (i.e. assumptions of Case 14 do not hold), the function γ is constant on closed sections of $\widetilde{\Omega}_v$, i.e. $\gamma(h_i) = \gamma(h_j)$ whenever h_i and h_j belong to the same closed section of $\widetilde{\Omega}_v$. Applying D 2, we can assume that every item in the section $[1, j+1]$ is covered exactly twice (i.e. $\gamma(h_i) = 2$ for every $i = 1, \ldots, j$) and for all $k \geq j+1$ we have $\gamma(h_k) > 2$. In this case we call the section $[1, j+1]$ *the quadratic part of Ω_v*. We sometimes refer to the set of non-quadratic sections of the generalised equation Ω_v as to the *non-quadratic part* of Ω_v.

Let $\mathrm{tp}(v) = 12$. Then the *quadratic part of Ω_v* is the whole active part of Ω_v.

A variable base μ of the generalised equation Ω is called a *quadratic base* if μ and its dual $\Delta(\mu)$ both belong to the quadratic part of Ω. A base μ of the generalised equation Ω is called a *quadratic-coefficient base* if μ belongs to the quadratic part of Ω and its dual $\Delta(\mu)$ does not belong to the quadratic part of Ω.

Let $\mathrm{tp}(v) = 12$ or $\mathrm{tp}(v) = 15$ and let $[1, j+1]$ be the quadratic part of Ω_v. An automorphism φ of the coordinate group $\mathbb{G}_{R(\Omega_v^*)}$ is called *invariant with respect to the non-quadratic part* if it is completely induced (see Definition 7.2) and its dual is invariant with respect to the non-quadratic part in the sense of Razborov, [**Raz87**]. We now recall the definition of an automorphism invariant with respect to the non-quadratic part in the sense of Razborov.

Let Υ_v be a non-constrained generalised equation over the free monoid $\mathbb{F}$, and $F_{R(\Upsilon_v^*)}$ be its coordinate group over F, where F is the free group with basis $\mathcal{A}$.

Denote by $\widehat{\Upsilon}_v$ the generalised equation obtained from Υ_v by removing all non-quadratic bases, all quadratic-coefficient bases and all coefficient equations. Let ϕ be the natural homomorphism from $F[h]/\mathrm{ncl}\langle\widehat{\Upsilon}_v\rangle$ to $F_{R(\Upsilon_v^*)}$.

An automorphism φ of the coordinate group $F_{R(\Upsilon_v^*)}$ is called *invariant with respect to the non-quadratic part*, see [**Raz87**], if there exists an automorphism $\hat{\varphi}$ of the group $F[h]/\mathrm{ncl}\langle\widehat{\Upsilon}_v\rangle$ so that for every quadratic-coefficient base ν of Υ_v one has $\hat{\varphi}(h(\nu)) = h(\nu)$, for every h_i, $i = j+1, \ldots, \rho_{\Upsilon_v}$ (recall that $[1, j+1]$ is the quadratic part of Υ_v) one has $\hat{\varphi}(h_i) = h_i$, and the following diagram commutes:

$$\begin{array}{ccc} F[h]/\mathrm{ncl}\langle\widehat{\Upsilon}_v\rangle & \xrightarrow{\phi} & F_{R(\Upsilon_v^*)} \\ \hat{\varphi}\downarrow & & \downarrow\varphi \\ F[h]/\mathrm{ncl}\langle\widehat{\Upsilon}_v\rangle & \xrightarrow{\phi} & F_{R(\Upsilon_v^*)} \end{array}$$

LEMMA 7.7 (Lemma 2.7, [**Raz87**]). *There exists a finitely presented subgroup K of the group of automorphisms of a free group such that every automorphism from K induces an automorphism of $F[h]/\mathrm{ncl}\langle\Upsilon_v\rangle$, which, in turn, induces an automorphism of the coordinate group $F_{R(\Upsilon_v^*)}$. The finitely generated group K' of all automorphisms of $F_{R(\Upsilon_v^*)}$ induced by automorphisms from K contains all the automorphisms invariant with respect to the non-quadratic part of Υ_v.*

We use the notation of Lemma 7.7. Every automorphism of $\mathbb{G}_{R(\Omega_v^*)}$ invariant with respect to the non-quadratic part is completely induced and, by definition, its dual belongs to K'. An argument analogous to the one in the case of automorphisms invariant with respect to the kernel shows that for every automorphism φ from K, there exists an algorithm to decide whether or not φ induces an automorphism of $\mathbb{G}_{R(\Omega_v^*)}$. We thereby get the following lemma.

LEMMA 7.8. *The group of all automorphisms of $\mathbb{G}_{R(\Omega_v^*)}$ invariant with respect to the non-quadratic part of Ω_v is tame and recursive.*

In this case, set $\mathfrak{V}(\Omega_v)$ to be the group of automorphisms of $\mathbb{G}_{R(\Omega_v^*)}$ invariant with respect to the non-quadratic part.

In all other cases set $\mathfrak{V}(\Omega_v) = 1$.

7.2. The finite subtree $T_0(\Omega)$

As mentioned above, the aim of this section is to construct the finite subtree $T_0(\Omega)$ of $T(\Omega)$ as the subtree that does not contain prohibited paths.

The definition of a prohibited path is designed in such a way that the paths $\mathfrak{p}(H)$ in the tree T associated to the solution H do not contain them. Therefore, the nature of the definition of a prohibited path will become clearer in Section 7.3.

By Lemma 4.19, infinite branches of the tree $T(\Omega)$ correspond to the following cases: $7 \leq \mathrm{tp}(v_k) \leq 10$ for all k, or $\mathrm{tp}(v_k) = 12$ for all k, or $\mathrm{tp}(v_k) = 15$ for all k. We now define prohibited paths of types 7-10, 12 and 15.

DEFINITION 7.9. *We call a path $v_1 \to v_2 \to \ldots \to v_k$ in $T(\Omega)$ prohibited of type 7-10 if $7 \leq \mathrm{tp}(v_i) \leq 10$ for all $i = 1, \ldots, k$ and some generalised equation with ρ variables occurs among $\{\Omega_{v_i} \mid 1 \leq i \leq l\}$ at least $2^{4\rho^2 \cdot (2^\rho + 1)} + 1$ times.*

Similarly, a path $v_1 \to v_2 \to \ldots \to v_k$ in $T(\Omega)$ is called *prohibited of type 12* if $\text{tp}(v_i) = 12$ for all $i = 1, \ldots, k$ and some generalised equation with ρ variables occurs among $\{\Omega_{v_i} \mid 1 \leq i \leq l\}$ at least $2^{4\rho^2 \cdot (2^\rho + 1)} + 1$ times.

We now prove that an infinite branch of $T(\Omega)$ of type 7-10 or 12 contains a prohibited path of type 7-10 or 12, correspondingly.

LEMMA 7.10. *Let $v_0 \to v_1 \to \ldots \to v_n \to \ldots$ be an infinite path in the tree $T(\Omega)$, where $7 \leq \text{tp}(v_i) \leq 10$ for all i, and let $\Omega_{v_0}, \Omega_{v_1}, \ldots, \Omega_{v_n}, \ldots$ be the sequence of corresponding generalised equations. Then among $\{\Omega_{v_i}\}$ some generalised equation occurs infinitely many times. Furthermore, if $\Omega_{v_k} = \Omega_{v_l}$, then $\pi(v_k, v_l)$ is a $\mathbb{G}$-automorphism of $\mathbb{G}_{R(\Omega_{v_k}^*)}$ invariant with respect to the kernel of Ω_{v_k}.*

PROOF. By Lemma 4.18, we have that $\text{comp}(\Omega_{v_k}) \leq \text{comp}(\Omega_{v_0})$ and $\xi(\Omega_{v_k}) \leq \xi(\Omega_{v_0})$ for all k. We, therefore, may assume that $\text{comp} = \text{comp}(\Omega_{v_k}) = \text{comp}(\Omega_{v_0})$ and $\xi(\Omega_{v_k}) = \xi(\Omega_{v_0})$ for all k. It follows that all the transformations ET 5 introduce a new boundary.

For all k, the generalised equations $\text{Ker}(\widetilde{\Upsilon}_{v_k})$ have the same set of bases, recall that $\widetilde{\Upsilon}_{v_k} = \text{D}\,3(\widetilde{\Upsilon}_{v_k})$. Indeed, consider the generalised equations $\widetilde{\Upsilon}_{v_k}$ and $\widetilde{\Upsilon}_{v_{k+1}}$. Since $\text{tp}(v_k) \neq 3, 4$, the active part of $\widetilde{\Upsilon}_{v_k}$ does not contain constant bases.

If $\text{tp}(v_k) = 7, 8, 10$, then $\widetilde{\Upsilon}_{v_{k+1}}$ is obtained from $\widetilde{\Upsilon}_{v_k}$ by cutting some base μ which is eliminable in $\widetilde{\Upsilon}_{v_k}$ and then deleting one of the new bases, which is also eliminable, since it falls under the assumption a) of the definition of an eliminable base. Since every transformation ET 5 introduces a new boundary, the remaining part of the base μ falls under the assumption b) of the definition of an eliminable base. Therefore, in this case, the set of bases that belong to the kernel does not change.

Let $\text{tp}(v_k) = 9$. It suffices to show that, in the notation of Case 9, all the bases of $\widetilde{\Upsilon}_{v_{k+1}}$ obtained by cutting the base μ_2 do not belong to the kernel. Without loss of generality we may assume that $\sigma(\mu_2)$ is a closed section of $\widetilde{\Upsilon}_{v_k}$. Indeed, if $\sigma(\mu_2)$ is not closed, instead, we can consider one of its closed subsections σ' in the generalised equation $\widetilde{\Upsilon}_{v_k}$.

Notice that, since $\sigma(\mu_2)$ is closed, every boundary that intersects μ_1 and μ_2 in Υ_{v_k}, touches exactly two bases in $\widetilde{\Upsilon}_{v_k}$. Thus, for every boundary connection (p, μ_2, q) in $\Upsilon_{v_{k+1}}$ either the boundary p or the boundary q touches exactly two bases in $\widetilde{\Upsilon}_{v_{k+1}}$.

Construct an elimination process (see description of the derived transformation D 4) for the generalised equation $\widetilde{\Upsilon}_{v_k}$ and take the first generalised equation Υ_i in this elimination process, such that the base ν eliminated in this equation was obtained from either μ_1 or μ_2 or $\Delta(\mu_1)$ or $\Delta(\mu_2)$ by applying D 3 to Υ_{v_k}. The base ν could not be obtained from μ_1 or μ_2, since every item in the section $\sigma(\mu_2)$ is covered twice and every boundary in this section touches two bases.

If ν falls under the assumption of case b) of the definition of an eliminable base, then

$$\text{either } \alpha(\nu) \in \{\alpha(\Delta(\mu_1)), \alpha(\Delta(\mu_2))\} \text{ or } \beta(\nu) \in \{\beta(\Delta(\mu_1)), \beta(\Delta(\mu_2))\}.$$

We now construct an elimination process for the generalised equation $\widetilde{\Upsilon}_{v_{k+1}}$. The first i steps of the elimination process for $\widetilde{\Upsilon}_{v_{k+1}}$ coincide with the first i steps of the elimination process constructed for the generalised equation $\widetilde{\Upsilon}_{v_k}$. Then the eliminable base ν of Υ_i corresponds to a base ν' obtained from either μ_2 or $\Delta(\mu_2)$ by applying D 3 to $\Upsilon_{v_{k+1}}$. The base ν' is eliminable.

Notice that one of the boundaries $\alpha(\nu)$, $\beta(\nu)$, $\alpha(\Delta(\nu))$ or $\alpha(\Delta(\nu))$ touches just two bases. Therefore, after eliminating ν', this boundary touches just one base η that was obtained from μ_2 or $\Delta(\mu_2)$. The base η falls under the assumptions b) of the definition of an eliminable base. Repeating this argument, one can subsequently eliminate all the other bases obtained from μ_2 or $\Delta(\mu_2)$. It follows that all the bases of the generalised equation $\widetilde{\Upsilon}_{v_{k+1}}$, obtained from μ_2 or $\Delta(\mu_2)$ do not belong to the kernel.

We thereby have shown that the set of bases is the same for all the generalised equations $\operatorname{Ker}(\widetilde{\Upsilon}_{v_k})$ and thus the set of bases is the same for all the generalised equations $\operatorname{Ker}(\widetilde{\Omega}_{v_k})$. We denote the cardinality of this set by n'.

We now prove that the number of bases in the active sections of Ω_{v_k}, for all k, is bounded above by a function of Ω_{v_0}:

(7.1) $$n_A(\Omega_{v_k}) \leq 3\operatorname{comp} + 6n' + 1.$$

Indeed, assume the contrary and let k be minimal for which inequality (4.5) fails. Then

(7.2) $$n_A(\Omega_{v_{k-1}}) \leq 3\operatorname{comp} + 6n' + 1, n_A(\Omega_{v_k}) > 3\operatorname{comp} + 6n' + 1.$$

By Lemma 4.18, $\operatorname{tp}(v_{k-1}) = 10$. It follows that $\operatorname{tp}(v_{k-1}) \neq 5, 6, 7, 8, 9$. Therefore, every active section of $\Omega_{v_{k-1}}$ either contains at least three bases or contains some base of the generalised equation $\operatorname{Ker}(\widetilde{\Omega}_{v_{k-1}})$. Let u_{k-1} and w_{k-1} be the number of active sections of $\Omega_{v_{k-1}}$ that contain one base and more than one base, respectively. Hence $u_{k-1} + w_{k-1} \leq \frac{1}{3} n_A(\Omega_{v_{k-1}}) + n'$. It is easy to see that

$$\operatorname{comp} = \sum_{\sigma \in A\Sigma(\Omega)} \max\{0, n(\sigma) - 2\} = n_A(\Omega_{v_{k-1}}) - 2w_{k-1} - u_{k-1}.$$

Then, $\operatorname{comp} \geq n_A(\Omega_{v_{k-1}}) - 2(w_{k-1} + u_{k-1}) \geq \frac{1}{3} n_A(\Omega_{v_{k-1}}) - 2n'$, which contradicts (7.2).

Furthermore, the number $\rho_A(\Omega_{v_k})$ of items in the active part of Ω_{v_k} is bounded above:

$$\rho_A(\Omega_{v_k}) \leq \xi(\Omega_{v_k}) + n_A(\Omega_{v_{k-1}}) + 1 \leq 3\operatorname{comp} + 6n' + \xi(\Omega_{v_0}) + 2.$$

Since the number of bases and number of items is bounded above, the set $\{\Omega_{v_k} \mid k \in \mathbb{N}\}$ is finite and thus some generalised equation occurs in this set infinitely many times.

Let $\Omega_{v_k} = \Omega_{v_l}$. Since, by assumption, the edges $v_j \to v_{j+1}$, $j = k, \ldots, l-1$ are labelled by isomorphisms, the homomorphism $\pi(v_k, v_l)$ is an automorphism of the coordinate group $\mathbb{G}_{R(\Omega_{v_k}^*)}$.

By Lemma 4.6 and Lemma 4.11, there exists an epimorphism $\pi(v_k, v_l)'$ from $F_{R(\Upsilon_{v_k}^*)}$ to $F_{R(\Upsilon_{v_l}^*)}$. Since $F_{R(\Upsilon_{v_k}^*)}$ is a finitely generated residually free group and therefore is residually finite, by a theorem of Mal'cev, $F_{R(\Upsilon_{v_k}^*)}$ is Hopfian. Thus the epimorphism $\pi(v_k, v_l)'$ is an automorphism of $F_{R(\Upsilon_{v_k}^*)}$. This shows that the automorphism $\pi(v_k, v_l)$ is completely induced and $\pi(v_k, v_l)'$ is its dual. We are left to show that $\pi(v_k, v_l)'$ is invariant with respect to the kernel (in the sense of Razborov, [**Raz87**]).

As shown above,
$$\mathrm{Ker}(\widetilde{\Upsilon}_{v_k}) = \mathrm{Ker}(\widetilde{\Upsilon}_{v_{k+1}}) = \cdots = \mathrm{Ker}(\widetilde{\Upsilon}_{v_l}).$$
From the above, it follows that $\mathrm{Ker}(\widetilde{\Upsilon}_{v_{i+1}})$ is obtained from $\mathrm{Ker}(\widetilde{\Upsilon}_{v_i})$ by introducing new boundaries and removing some of the items that do not belong to the kernel of $\widetilde{\Upsilon}_{v_{i+1}}$. Therefore, the number of items that belong to the kernel $\mathrm{Ker}(\widetilde{\Upsilon}_{v_{i+1}})$ can only increase. As $\Omega_{v_k} = \Omega_{v_l}$, so this number is the same for all i, $i = k, \ldots, l-1$. It follows that $\pi(v_k, v_l)'(h_i) = h_i$ for all h_i that belong to the kernel of Υ_{v_k}.

Since the transformations that take the generalised equation $\widetilde{\Upsilon}_{v_k}$ to $\widetilde{\Upsilon}_{v_{k+1}}$ do not involve bases that belong to the kernel of $\widetilde{\Upsilon}_{v_k}$, the same sequence of transformations can be applied to the generalised equation $\widehat{\widetilde{\Upsilon}}_{v_k}$, where $\widehat{\widetilde{\Upsilon}}_{v_k}$ is obtained from $\widetilde{\Upsilon}_{v_k}$ by removing all coefficient equations and all bases that belong to the kernel of $\widetilde{\Upsilon}_{v_k}$.

Since, by assumption, every time we μ-tie a boundary a new boundary is introduced, we get that the epimorphism from $F_{R(\widehat{\widetilde{\Upsilon}}^*_{v_k})}$ to $F_{R(\widehat{\widetilde{\Upsilon}}^*_{v_{k+1}})}$ is, in fact, an isomorphism. We therefore get the following commutative diagram (see the definition of an automorphism invariant with respect to the kernel, Section 7.1):

$$\begin{array}{ccc} F_{R(\widehat{\Upsilon}^*_{v_k})} & \xrightarrow{\pi} & F_{R(\Upsilon_{v_k}^*)} \\ \downarrow & & \downarrow \pi(v_k,v_l)' \\ F_{R(\widehat{\Upsilon}^*_{v_l})} & \xrightarrow{\pi} & F_{R(\Upsilon_{v_l}^*)} \end{array}$$

It follows that the automorphism $\pi(v_k, v_l)'$ is invariant with respect to the kernel in the sense of Razborov and thus the automorphism $\pi(v_k, v_l)$ is invariant with respect to the kernel of Ω_{v_k}. $\square$

COROLLARY 7.11. *Let* $\mathfrak{p} = v_1 \to \cdots \to v_n \to \cdots$ *be an infinite path in the tree* $T(\Omega)$, *and* $7 \leq \mathrm{tp}(v_i) \leq 10$ *for all* i. *Then* $\mathfrak{p}$ *contains a prohibited subpath of type 7-10.*

LEMMA 7.12. *Let* $v_0 \to v_1 \to \cdots \to v_n \to \cdots$ *be an infinite path in the tree* $T(\Omega)$, *where* $\mathrm{tp}(v_i) = 12$ *for all* i, *and* $\Omega_{v_0}, \Omega_{v_1}, \ldots, \Omega_{v_n}, \ldots$ *be the sequence of corresponding generalised equations. Then among* $\{\Omega_{v_i}\}$ *some generalised equation occurs infinitely many times. Furthermore, if* $\Omega_{v_k} = \Omega_{v_l}$, *then* $\pi(v_k, v_l)$ *is a* $\mathbb{G}$-*automorphism of the coordinate group* $\mathbb{G}_{R(\Omega^*_{v_k})}$ *invariant with respect to the non-quadratic part.*

PROOF. Notice that since Ω_{v_i} is a quadratic generalised equation, quadratic-coefficient bases of Ω_{v_i} are bases whose duals belong to the non-active part.

Let μ_i be the carrier base of the generalised equation Ω_{v_i}. Consider the sequence $\mu_0, \ldots, \mu_i, \ldots$ of carrier bases. By Lemma 4.18, if $\mathrm{tp}(v_i) = 12$, then $n_A(\Omega_{v_{i+1}}) \leq n_A(\Omega_{v_i})$. Furthermore, if the carrier base μ_i is quadratic-coefficient, then this inequality is strict. Hence, it suffices to consider the case when all carrier bases are quadratic.

The number of consecutive quadratic bases in the sequence $\mu_1, \ldots, \mu_i, \ldots$ is bounded above. Indeed, by Lemma 4.18, when the entire transformation is applied, the complexity of the generalised equation does not increase. Furthermore, since the generalised equation is quadratic and does not contain free boundaries, the number of items does not increase. The number of constrained generalised equations with a bounded number of items and bounded complexity is finite and thus some generalised equation occurs in the sequence $\{\Omega_{v_i}\}$ infinitely many times.

Obviously, if $\Omega_{v_k} = \Omega_{v_l}$, then $\pi(v_k, v_l)$ is a $\mathbb{G}$-automorphism of $\mathbb{G}_{R(\Omega_{v_k}^*)}$.

By Lemma 4.6 and Lemma 4.11, there exists an epimorphism $\pi(v_k, v_l)'$ from $F_{R(\Upsilon_{v_k}^*)}$ to $F_{R(\Upsilon_{v_l}^*)}$. Since $F_{R(\Upsilon_{v_k}^*)}$ is a finitely generated residually free group and therefore is residually finite, by a theorem of Mal'cev, $F_{R(\Upsilon_{v_k}^*)}$ is Hopfian. Thus the epimorphism $\pi(v_k, v_l)'$ is an automorphism of $F_{R(\Upsilon_{v_k}^*)}$. This shows that the automorphism $\pi(v_k, v_l)$ is completely induced and $\pi(v_k, v_l)'$ is its dual. We are left to show that $\pi(v_k, v_l)'$ is invariant with respect to the non-quadratic part (in the sense of Razborov, [**Raz87**]).

From the definition of the entire transformation D 5, it follows that the number of items that belong to a given quadratic-coefficient base can only increase. As $\Omega_{v_k} = \Omega_{v_l}$, so this number is the same for all i, $i = k, \ldots, l$. It follows that $\pi(v_k, v_l)'(h_i) = h_i$ for all h_i that belong to a quadratic-coefficient base.

Since the transformations that take the generalised equation $\widetilde{\Upsilon}_{v_k}$ to $\widetilde{\Upsilon}_{v_{k+1}}$ involve only quadratic bases of $\widetilde{\Upsilon}_{v_k}$, the same sequence of transformations can be applied to the generalised equation $\widehat{\widetilde{\Upsilon}}_{v_k}$, where $\widehat{\widetilde{\Upsilon}}_{v_k}$ is the generalised equation obtained from $\widetilde{\Upsilon}_{v_k}$ by removing all non-quadratic bases, all quadratic-coefficient bases and all coefficient equations.

Since, by assumption, every time we μ-tie a boundary a new boundary is introduced, we get that the epimorphism from $F_{R(\widehat{\widetilde{\Upsilon}}_{v_k}^*)}$ to $F_{R(\widehat{\widetilde{\Upsilon}}_{v_{k+1}}^*)}$ is, in fact, an isomorphism. We therefore get the following commutative diagram (see the definition of an automorphism invariant with respect to the non-quadratic part, Section 7.1):

$$\begin{array}{ccc} F[h]/\mathrm{ncl}\langle\widehat{\Upsilon}_{v_k}\rangle & \xrightarrow{\phi} & F_{R(\Upsilon_{v_k}^*)} \\ \downarrow & & \downarrow \pi(v_k,v_l)' \\ F[h]/\mathrm{ncl}\langle\widehat{\Upsilon}_{v_l}\rangle & \xrightarrow{\phi} & F_{R(\Upsilon_{v_l}^*)} \end{array}$$

It follows that the automorphism $\pi(v_k, v_l)'$ is invariant with respect to the non-quadratic part of Υ_{v_k} in the sense of Razborov and thus the automorphism $\pi(v_k, v_l)$ is invariant with respect to the non-quadratic part of Ω_{v_k}. □

COROLLARY 7.13. *Let* $\mathfrak{p} = v_1 \to \ldots \to v_n \to \ldots$ *be an infinite path in the tree* $T(\Omega)$, *and* $\mathrm{tp}(v_i) = 12$ *for all* i. *Then* $\mathfrak{p}$ *contains a prohibited subpath of type 12.*

Below we shall define prohibited paths of type 15 in $T(\Omega)$. In this case, the definition of a prohibited path is much more involved.

We need some auxiliary definitions. Recall that the complexity of a generalised equation Ω is defined as follows:
$$\mathrm{comp} = \mathrm{comp}(\Omega) = \sum_{\sigma \in A\Sigma_\Omega} \max\{0, n(\sigma) - 2\},$$
where $n(\sigma)$ is the number of bases in σ. Let $\tau_v = \tau(\Omega_v) = \mathrm{comp}(\Omega_v) + \rho - \rho'_v$, where $\rho = \rho_\Omega$ is the number of variables in the initial generalised equation Ω and ρ'_v is the number of free variables belonging to the non-active sections of the generalised

equation Ω_v. We have $\rho'_v \le \rho$ (see the proof of Lemma 4.19), hence $\tau_v \ge 0$. If, in addition, $v_1 \to v_2$ is an auxiliary edge, then $\tau_{v_2} < \tau_{v_1}$.

We use induction on τ_v to construct a finite subtree $T_0(\Omega_v)$ of $T(\Omega_v)$, and the function $\mathbf{s}(\Omega_v)$.

The tree $T_0(\Omega_v)$ is a rooted tree at v and consists of some of the vertices and edges of $T(\Omega)$ that lie above v.

Suppose that $\tau_v = 0$. It follows that $\mathrm{comp}(\Omega_v) = 0$. Then in $T(\Omega)$ there are no auxiliary edges. Furthermore, since $\mathrm{comp}(\Omega_v) = 0$, it follows that every closed active section contains at most two bases and so no vertex of type 15 lies above v. We define the subtree $T_0(\Omega_v)$ as follows. The set of vertices of $T_0(\Omega_v)$ consists of all vertices v_1 of $T(\Omega)$ that lie above v, and so that the path from v to v_1 does not contain prohibited subpaths of types 7-10 and 12. By Corollary 7.11 and Corollary 7.13, $T_0(\Omega_v)$ is finite.

Let

(7.3) $$\mathbf{s}(\Omega_v) = \max_w \max_{\langle \mathcal{P}, R \rangle} \rho_{\Omega_w} \cdot \{f_0(\Omega_w, \mathcal{P}, R)\},$$

where $\max_w$ is taken over all the vertices of $T_0(\Omega_v)$ for which $\mathrm{tp}(w) = 2$ and Ω_w contains non-active sections; $\max_{\langle \mathcal{P}, R \rangle}$ is taken over all regular periodic structures such that the generalised equation $\widetilde{\Omega}_w$ is regular with respect to $\langle \mathcal{P}, R \rangle$; and f_0 is the function from Lemma 6.20.

Suppose now that $\tau_v > 0$. By induction, we assume that for all v_1 such that $\tau_{v_1} < \tau_v$ the finite tree $T_0(\Omega_{v_1})$ and $\mathbf{s}(\Omega_{v_1})$ are already defined. Furthermore, we assume that the full subtree of $T(\Omega)$ whose set of vertices consists of all vertices that lie above v does not contain prohibited paths of type 7-10 and of type 12. Consider a path $\mathfrak{p}$ in $T(\Omega)$:

(7.4) $$v_1 \to v_2 \to \ldots \to v_m,$$

where $\mathrm{tp}(v_i) = 15$, $1 \le i \le m$ and all the edges are principal. We have $\tau_{v_i} = \tau_v$.

Denote by μ_i the carrier base of the generalised equation Ω_{v_i}. Path (7.4) is called μ-reducing if $\mu_1 = \mu$ and either there are no auxiliary edges from the vertex v_2 and μ occurs in the sequence $\mu_1, \ldots, \mu_{m-1}$ at least twice, or there are auxiliary edges $v_2 \to w_1, v_2 \to w_2, \ldots, v_2 \to w_n$ and μ occurs in the sequence $\mu_1, \ldots, \mu_{m-1}$ at least $\max_{1 \le i \le n} \mathbf{s}(\Omega_{w_i})$ times. We will show later, see Equation (7.19), that, informally, in any μ-reducing path the length of the solution H is reduced by at least $\frac{1}{10}$ of the length of $H(\mu)$, hence the terminology.

DEFINITION 7.14. Path (7.4) is called *prohibited of type 15*, if it can be represented in the form

(7.5) $$\mathfrak{p}_1 \mathfrak{s}_1 \ldots \mathfrak{p}_l \mathfrak{s}_l \mathfrak{p}',$$

where for some sequence of bases $\eta_1, \ldots, \eta_l$ the following three conditions are satisfied:

(1) the path $\mathfrak{p}_i$ is η_i-reducing;
(2) every base μ_i that occurs at least once in the sequence $\mu_1, \ldots, \mu_{m-1}$, occurs at least $40n^2 f_1(\Omega_{v_2}) + 20n + 1$ times in the sequence $\eta_1, \ldots, \eta_l$, where $n = |\mathcal{BS}(\Omega_{v_i})|$ is the number of all bases in the generalised equation Ω_{v_i}, and f_1 is the function from Lemma 7.19; in other words, in a prohibited

path of type 15, for every carrier base μ_i there exists at least $40n^2 f_1(\Omega_{v_2}) + 20n + 1$ many μ_i-reducing paths.

(3) every transfer base of some generalised equation of the path $\mathfrak{p}$ is a transfer base of some generalised equation of the path $\mathfrak{p}'$.

Note that for any path of the form (7.4) in $T(\Omega)$, there is an algorithm to decide whether this path is prohibited of type 15 or not.

We now prove that any infinite branch of the tree $T(\Omega)$ of type 15 contains a prohibited subpath of type 15.

LEMMA 7.15. *Let* $\mathfrak{p} = v_1 \to \ldots \to v_n \to \ldots$ *be an infinite path in the tree* $T(\Omega)$, *and* $\mathrm{tp}(v_i) = 15$ *for all* i. *Then* $\mathfrak{p}$ *contains a prohibited subpath of type 15.*

PROOF. Let ω be the set of all bases occurring in the sequence of carrier bases $\mu_1, \mu_2, \ldots$ infinitely many times, and $\tilde{\omega}$ be the set of all bases that are transfer bases of infinitely many equations Ω_{v_i}. Considering, if necessary, a subpath $\tilde{\mathfrak{p}}$ of $\mathfrak{p}$ of the form $v_j \to v_{j+1} \to \ldots$, one can assume that all the bases in the sequence $\mu_1, \mu_2 \ldots$ belong to ω and every base which is a transfer base of at least one generalised equation belongs to $\tilde{\omega}$. Then for any $\mu \in \omega$ the path $\tilde{\mathfrak{p}}$ contains infinitely many non-intersecting μ-reducing finite subpaths. Hence there exists a subpath of the form (7.5) of $\tilde{\mathfrak{p}}$ which satisfies conditions (1) and (2) of the definition of a prohibited path of type 15, see Definition 7.14. Taking a long enough subpath $\mathfrak{p}'$ of $\mathfrak{p}$, we obtain a prohibited subpath of $\mathfrak{p}$. □

We now construct the tree $T_0(\Omega)$. Let $T'(\Omega_v)$ be the subtree of $T(\Omega_v)$ consisting of the vertices v_1 such that the path from v to v_1 in $T(\Omega)$ does not contain prohibited subpaths and does not contain vertices $v_2 \neq v_1$ such that $\tau_{v_2} < \tau_v$. Thus, the leaves of $T'(\Omega_v)$ are either vertices v_1 such that $\tau_{v_1} < \tau_v$ or leaves of $T(\Omega_v)$.

The subtree $T'(\Omega_v)$ can be effectively constructed. The tree $T_0(\Omega_v)$ is obtained from $T'(\Omega_v)$ by attaching (gluing) $T_0(\Omega_{v_1})$ (which is already constructed by the induction hypothesis) to those leaves v_1 of $T'(\Omega_v)$ for which $\tau_{v_1} < \tau_v$. The function $\mathbf{s}(\Omega_v)$ is defined by (7.3). Set $T_0(\Omega) = T_0(\Omega_{v_0})$, which is finite by construction.

DEFINITION 7.16. Denote by $\mathrm{Aut}(\Omega)$, $\Omega = \Omega_{v_0}$, the group of automorphisms of $\mathbb{G}_{R(\Omega^*)}$, generated by all the groups $\pi(v_0, v) \mathfrak{V}(\Omega_v) \pi(v_0, v)^{-1}$, $v \in T_0(\Omega)$, $\mathrm{tp}(v) \neq 1$ (thus $\pi(v_0, v)$ is an isomorphism). Note that by construction the group $\mathrm{Aut}(\Omega)$ is recursive.

We adopt the following convention. Given two solutions $H^{(i)}$ of Ω_{v_i} and $H^{(i')}$ of $\Omega_{v'_i}$, by $H^{(i)} <_{\mathrm{Aut}(\Omega)} H^{(i')}$ we mean $H^{(i)} <_{\pi(v_0, v_{i'})^{-1} \mathrm{Aut}(\Omega) \pi(v_0, v_i)} H^{(i')}$.

LEMMA 7.17. *Let*

$$(\Omega, H) = (\Omega_{v_0}, H^{(0)}) \to (\Omega_{v_1}, H^{(1)}) \to \ldots \to (\Omega_{v_l}, H^{(l)})$$

be the path defined by the solution H. *If* H *is a minimal solution with respect to the group of automorphisms* $\mathrm{Aut}(\Omega)$, *then* $H^{(i)}$ *is a minimal solution of* Ω_{v_i} *with respect to the group* $\mathfrak{V}(\Omega_{v_i})$ *for all* i.

PROOF. Follows from Lemma 5.8. □

7.3. Paths $\mathfrak{p}(H)$ are in $T_0(\Omega)$

The goal of this section is to give a proof of the proposition below.

PROPOSITION 7.18. *For any solution H of a generalised equation Ω there exists a leaf w of the tree $T_0(\Omega)$, $\mathrm{tp}(w) = 1, 2$, and a solution $H^{[w]}$ of the generalised equation Ω_w such that*

(1) $H^{[w]} <_{\mathrm{Aut}(\Omega)} H$;
(2) *if $\mathrm{tp}(w) = 2$ and the generalised equation Ω_w contains non-constant non-active sections, then there exists a period P such that $H^{[w]}$ is periodic with respect to the period P, and the generalised equation Ω_w is singular or strongly singular with respect to the periodic structure $\mathcal{P}(H^{[w]}, P)$.*

The proof of this proposition is rather long and technical. We now outline the organisation of the proof.

In part (A), for any solution H of Ω we describe a path $\mathfrak{p}(H) : (\Omega_{v_0}, H^{[0]}) \to \cdots \to (\Omega_{v_l}, H^{[l]})$, where $H^{[i]} <_{\mathrm{Aut}(\Omega)} H$.

In part (B) we prove that all vertices of the paths $\mathfrak{p}(H)$ belong to the tree T_0. In order to do so we show that they do not contain prohibited subpaths. In steps (I) and (II) we prove that the paths $\mathfrak{p}(H)$ do not contain prohibited paths of type 7-10 and 12, correspondingly. The proof is by contradiction: if $\mathfrak{p}(H)$ is not in T_0, the fact that a generalised equation repeats enough times, allows us to construct an automorphism that makes the solution shorter, contradicting its minimality.

To prove that the paths $\mathfrak{p}(H)$ do not contain prohibited paths of type 15 (step (III)) we show, by contradiction, that on the one hand, the length of minimal solutions is bounded above by a function of the excess (see Definition 4.8 for definition of excess), see Equation (7.15) and, on the other hand the length of a minimal solution H such that $\mathfrak{p}(H)$ contains a prohibited path of type 15 fails inequality (7.15).

Finally, in part (C) we prove that the pair $(\Omega_w, H^{[w]})$, where w is a leaf of type 2, satisfies the properties required in Proposition 7.18.

(A): **Constructing the paths $\mathfrak{p}(H)$.** To define the path $\mathfrak{p}(H)$ we shall make use of two functions $\mathbf{e}$ and $\mathbf{e}'$ that assign to the pair $(\Omega_v, H^{(v)})$ a pair $(\Omega_{v'}, H^{(v')})$, where either $v' = v$ or there is an edge $v \to v'$ in $T(\Omega)$. The function $\mathbf{e}$ can be applied to any pair $(\Omega_v, H^{(v)})$, where $\mathrm{tp}(v) \neq 1, 2$ and the function $\mathbf{e}'$ can only be applied to a pair $(\Omega_v, H^{(v)})$, where $\mathrm{tp}(v) = 15$ and there are auxiliary edges outgoing from the vertex v.

We now define the functions $\mathbf{e}$ and $\mathbf{e}'$.

Let $\mathrm{tp}(v) = 3$ or $\mathrm{tp}(v) \geq 6$ and let $v \to w_1, \ldots, v \to w_m$ be the list of all *principal* outgoing edges from v, then the generalised equations $\Omega_{w_1}, \ldots, \Omega_{w_m}$ are obtained from Ω_v by a sequence of elementary transformations. For every solution H the path defined by H is unique, i.e. for the pair (Ω, H) there exists a unique pair $(\Omega_{w_i}, H^{(i)})$ such that the following diagram commutes:

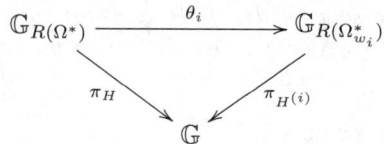

Define a function **e** that assigns the pair $(\Omega_{w_i}, H^{(i)})$ to the pair (Ω_v, H), $\mathbf{e}(\Omega_v, H) = (\Omega_{w_i}, H^{(i)})$.

Let $\mathrm{tp}(v) = 4$ or $\mathrm{tp}(v) = 5$. In these cases there is a single edge $v \to w_1$ outgoing from v and this edge is auxiliary. We set $\mathbf{e}(\Omega_v, H) = (\Omega_{w_1}, H^{(1)})$.

If $\mathrm{tp}(v) = 15$ and there are auxiliary outgoing edges from the vertex v, then the carrier base μ of the generalised equation Ω_v intersects with $\Delta(\mu)$. Below we use the notation from the description of Case 15.1. For any solution H of the generalised equation Ω_v one can construct a solution H' of the generalised equation $\Omega_{v'}$ as follows: $H'_{\rho_v+1} = H[1, \beta(\Delta(\mu))]$. We define the function $\mathbf{e}'$ as follows $\mathbf{e}'(\Omega_v, H) = (\Omega_{v'}, H')$.

To construct the path $\mathfrak{p}(H)$

$$(7.6) \qquad (\Omega, H) \to (\Omega_{v_0}, H^{[0]}) \to (\Omega_{v_1}, H^{[1]}) \to \ldots$$

we use induction on its length i.

Let $i = 0$, we define $H^{[0]}$ to be a solution of the generalised equation Ω minimal with respect to the group of automorphisms $\mathrm{Aut}(\Omega)$, such that $H^{[0]} <_{\mathrm{Aut}(\Omega)} H$. Let $i \geq 1$ and suppose that the term $(\Omega_{v_i}, H^{[i]})$ of the sequence (7.6) is already constructed. We construct $(\Omega_{v_{i+1}}, H^{[i+1]})$

If $3 \leq \mathrm{tp}(v_i) \leq 6$, $\mathrm{tp}(v_i) = 11, 13, 14$, we set $(\Omega_{v_{i+1}}, H^{[i+1]}) = \mathbf{e}(\Omega_{v_i}, H^{[i]})$.

If $7 \leq \mathrm{tp}(v_i) \leq 10$ or $\mathrm{tp}(v_i) = 12$ and there exists a minimal solution H^+ of Ω_{v_i} such that $H^+ <_{\mathrm{Aut}(\Omega)} H^{[i]}$ and $|H^+| < |H^{[i]}|$, then we set $(\Omega_{v_{i+1}}, H^{[i+1]}) = (\Omega_{v_i}, H^+)$. Note that, since $H^{[0]}$ is a minimal solution of Ω with respect to the group of automorphisms $\mathrm{Aut}(\Omega)$, by construction and by Lemma 7.17, we have that the solution $H^{[i]}$ is minimal with respect to the group of automorphism $\mathfrak{V}(\Omega_{v_i})$ for all i. Although $H^{[i]}$ is a minimal solution, in this step we take a minimal solution of minimal total length, see Remark 5.7.

Let $\mathrm{tp}(v_i) = 15$, $v_i \neq v_{i-1}$ and $v_i \to w_1, \ldots, v_i \to w_\mathbf{n}$ be the auxiliary edges outgoing from v_i (the carrier base μ intersects with its dual $\Delta(\mu)$). If there exists a period P such that

$$(7.7) \qquad H^{[i]}[1, \beta(\Delta(\mu))] \doteq P^r P_1, \ P \doteq P_1 P_2, \ r \geq \max_{1 \leq i \leq \mathbf{n}} \mathbf{s}(\Omega_{w_i}),$$

then we set $(\Omega_{v_{i+1}}, H^{[i+1]}) = \mathbf{e}'(\Omega_{v_i}, H^{[i]})$ and declare the section $[1, \beta(\Delta(\mu))]$ non-active.

In all the other cases (when $\mathrm{tp}(v_i) = 15$) we set $(\Omega_{v_{i+1}}, H^{[i+1]}) = \mathbf{e}(\Omega_{v_i}, H^{[i]})$.

The path (7.6) ends if $\mathrm{tp}(v_i) \leq 2$.

A leaf w of the tree $T(\Omega)$ is called *final* if there exists a solution H of Ω_{v_0} and a path $\mathfrak{p}(H)$ such that $\mathfrak{p}(H)$ ends in w.

(B): **Paths $\mathfrak{p}(H)$ belong to T_0.** We use induction on τ to show that every vertex v_i of the path $\mathfrak{p}(H)$ (see Equation (7.6)) belongs to $T_0(\Omega)$, i.e. $v_i \in T_0(\Omega)$. Suppose that $v_i \notin T_0(\Omega)$ and let i_0 be the least among such numbers. It follows from the construction of $T_0(\Omega)$ that there exists $i_1 < i_0$ such that the path from v_{i_1} to v_{i_0} contains a prohibited subpath $\mathfrak{s}$. From the minimality of i_0 it follows that the prohibited path $\mathfrak{s}$ goes from v_{i_2}, $i_1 \leq i_2 \leq i_0$ to v_{i_0}.

(I): **Paths $\mathfrak{p}(H)$ do not contain prohibited subpaths of type 7-10.** Suppose first that the prohibited path $\mathfrak{s}$ is of type 7-10, i.e. $7 \leq \mathrm{tp}(v_i) \leq 10$. By definition, there exists a generalised equation $\Omega_{v_{k_1}}$ that repeats $r = 2^{4\rho_{\Omega_{v_{k_1}}}^2 \cdot (2^{\rho_{\Omega_{v_{k_1}}}}+1)} + 1$ times, i.e.

$$\langle \Upsilon_{v_{k_1}}, \Re\Upsilon_{v_{k_1}} \rangle = \cdots = \langle \Upsilon_{v_{k_r}}, \Re\Upsilon_{v_{k_r}} \rangle.$$

Since the path $\mathfrak{s}$ is prohibited, we may assume that $v_{k_i} \neq v_{k_i+1}$ for all i and $v_{k_i} \neq v_{k_i+1}$ for all i, i.e. $(\Omega_{v_{k_i+1}}, H^{[k_i+1]}) = \mathbf{e}(\Omega_{v_{k_i}}, H^{[k_i]})$.

We now prove that there exist k_j and $k_{j'}$, $k_j < k_{j'}$ such that $H^{[k_{j'}]} <_{\mathrm{Aut}(\Omega)} H^{[k_j]}$.

Since, the number of different $2\rho_{\Omega_{v_{k_1}}} \times 2\rho_{\Omega_{v_{k_1}}}$ cancellation matrices (see Definition 5.4) is bounded above by r, if the generalised equation $\Omega_{v_{k_1}}$ repeats r times, there exist k_j and $k_{j'}$ such that $H^{[k_j]}$ and $H^{[k_{j'}]}$ have the same cancellation matrix, i.e. satisfy conditions (2) and (3) from Definition 5.1. Moreover, by Lemma 7.10, $\pi(v_{k_j}, v_{k_{j'}})$ is an automorphism of $\mathbb{G}_{R(\Omega_{v_{k_j}}^*)}$ invariant with respect to the kernel of $\Omega_{v_{k_j}}$.

By Remark 4.20, we have $|H^{[k_j]}| > |H^{[k_{j'}]}|$. This derives a contradiction, since, by construction of the sequence (7.6) one has $v_{k_j+1} = v_{k_j}$.

(II): **Paths $\mathfrak{p}(H)$ do not contain prohibited subpaths of type 12.** Suppose next that the path $\mathfrak{s}$ is prohibited of type 12, i.e. $\mathrm{tp}(v_i) = 12$. An analogous argument to the one for prohibited paths of type 7-10, but using Lemma 7.12 instead of Lemma 7.10, leads to a contradiction. Hence, we conclude that $v_i \in T_0(\Omega)$, where $\mathrm{tp}(v_i) = 12$.

(III): **Paths $\mathfrak{p}(H)$ do contain prohibited subpaths of type 15.** Finally, suppose that the path $\mathfrak{s}$ is prohibited of type 15, i.e. $\mathrm{tp}(v_i) = 15$. Abusing the notation, we consider a subpath of (7.6)

$$(\Omega_{v_1}, H^{[1]}) \to (\Omega_{v_2}, H^{[2]}) \to \cdots \to (\Omega_{v_m}, H^{[m]}) \to \cdots,$$

where $v_1, v_2, \ldots$ are vertices of the tree $T_0(\Omega)$, $\mathrm{tp}(v_i) = 15$ and the edges $v_i \to v_{i+1}$ are principal for all i. Notice, that by construction the above path is the path defined by the solution $H^{(1)} = H^{[1]}$:

(7.8) $$(\Omega_{v_1}, H^{(1)}) \to (\Omega_{v_2}, H^{(2)}) \to \cdots \to (\Omega_{v_m}, H^{(m)}) \to \cdots,$$

To simplify the notation, below we write ρ_i for $\rho_{\Omega_{v_i}}$.

Let $\omega = \{\mu_1, \ldots, \mu_m, \ldots\}$ be the set of carrier bases μ_i of the generalised equations Ω_{v_i}'s and let $\tilde{\omega}$ denote the set of bases which are transfer bases for at least one generalised equation in (7.8). By ω_2 we denote the set of all bases ν of Ω_{v_i}, $i = 1, \ldots, m, \ldots$ so that $\nu, \Delta(\nu) \notin \omega \cup \tilde{\omega}$. Let

$$\alpha(\omega) = \min\left\{\min_{\mu \in \omega_2}\{\alpha(\mu)\}, \rho_A\right\},$$

where ρ_A is the boundary between the active part and the non-active part.

For every element $(\Omega_{v_i}, H^{(i)})$ of the sequence (7.8), using D 3, if necessary, we make the section $[1, \alpha(\omega)]$ of the generalised equation Ω_{v_i} closed and set $[\alpha(\omega), \rho_i]$ to be the non-active part of the generalised equation Ω_{v_i} for all i.

Recall that by ω_1 we denote the set of all variable bases ν for which either ν or $\Delta(\nu)$ belongs to the active part $[1, \alpha(\omega)]$ of the generalised equation Ω_{v_1}, see Definition 4.8.

(III.1): Lengths of minimal solutions are bounded by a function of the excess. Let H be a solution of the generalised equation Ω and let $[1, j+1]$ be the quadratic part of Ω. Set

$$d_1(H) = \sum_{i=1}^{j} |H_i|, \ d_2(H) = \sum_{\nu} |H(\nu)|,$$

where ν is a quadratic-coefficient base.

LEMMA 7.19. *Let v be a vertex of $T(\Omega)$, $\mathrm{tp}(v) = 15$. There exists a recursive function $f_1(\Omega_v)$ such that for any solution H minimal with respect to $\mathfrak{V}(\Omega_v)$ one has*

$$d_1(H) \le f_1(\Omega_v) \max\{d_2(H), 1\}.$$

PROOF. Since $\mathrm{tp}(v) = 15$, every boundary that touches a base is η-tied in every base η which it intersects. Instead of Ω_v below we consider the generalised equation $\widetilde{\Omega}_v = \mathrm{D}3(\Omega_v)$ which does not have any boundary connections. Then $\mathbb{G}_{R(\Omega_v^*)}$ is isomorphic to $\mathbb{G}_{R(\widetilde{\Omega}_v^*)}$. We abuse the notation and denote $\widetilde{\Omega}_v$ by Ω_v.

Consider the sequence

$$(\Omega_v, H) = (\Omega_{v_0}, H^{(0)}) \to (\Omega_{v_1}, H^{(1)}) \to \cdots \to (\Omega_{v_i}, H^{(i)}) \to \cdots,$$

where $(\Omega_{v_{j+1}}, H^{(j+1)})$ is obtained from $(\Omega_{v_j}, H^{(j)})$ by applying the entire transformation D5 in the quadratic part of Ω. Denote by μ_i the carrier base of the generalised equation Ω_{v_i} and consider the sequence $\mu_1, \ldots, \mu_i, \ldots$

We use an argument analogous to the one given in the proof of Lemma 7.12 to show that the number of consecutive quadratic bases in the sequence $\mu_1, \ldots, \mu_i, \ldots$ is bounded above. The entire transformation applied in the quadratic part of a generalised equation, does not increase the complexity and the number of items. The number of constrained generalised equations with a bounded number of items and bounded complexity is finite.

We now prove that if a generalised equation $\Omega_{v_{k_1}}$ repeats $r = 2^{4\rho_{k_1}^2 \cdot (2_{k_1}^\rho + 1)} + 1$ times, then there exist k_j and $k_{j'}$ such that $H^{(k_{j'})} <_{\mathfrak{V}(\Omega_{v_{k_j}})} H^{(k_j)}$.

Since, the number of different $2\rho_{k_1} \times 2\rho_{k_1}$ cancellation matrices (see Definition 5.4) is bounded above by r, if the generalised equation $\Omega_{v_{k_1}}$ repeats r times, there exist k_j and $k_{j'}$ such that $H^{(k_j)}$ and $H^{(k_{j'})}$ have the same cancellation matrix, i.e. satisfy conditions (2) and (3) from Definition 5.1.

Moreover, by Lemma 7.12, the automorphism $\pi(v_{k_j}, v_{k_{j'}})$ of $\mathbb{G}_{R(\Omega_{v_{k_j}}^*)}$ is invariant with respect to the non-quadratic part of $\Omega_{v_{k_j}}$.

By Remark 4.20, we have $|H^{(k_{j'})}| < |H^{(k_j)}|$. We, therefore, have that $H^{(k_{j'})} <_{\mathfrak{V}(\Omega_{v_{k_j}})} H^{(k_j)}$, contradicting the minimality of $H^{(k_j)}$. Thus, we have proven that the number of consecutive quadratic carrier bases is bounded.

Furthermore, since whenever the carrier base is a quadratic-coefficient base the number of bases in the quadratic part decreases, there exists an integer N bounded above by a computable function of the generalised equation, such that the quadratic part of Ω_{v_N} is empty.

We prove the statement of the lemma by induction on the length i of the sequence. If $i = N - 1$, since the application of the entire transformation to $\Omega_{v_{N-1}}$ results in a generalised equation with the empty quadratic part, the carrier μ_{N-1} is a quadratic-coefficient base and all the other bases $\nu_{N-1,1}, \ldots, \nu_{N-1,n_{N-1}}$ are transfer bases and are transferred from the carrier to its dual. Since the length $|H^{(N-1)}(\nu_{N-1,i})|$ of every transferred base is less than the length of the carrier, we get that

$$d_1(H^{(N-1)}) \leq \sum_{n=1}^{n_{N-1}} |H^{(N-1)}(\nu_{N-1,n})| \leq n_{N-1}|H^{(N-1)}(\mu_{N-1})| = n_{N-1}d_2(H^{(N-1)}).$$

By induction, we may assume that

$$d_1(H^{(i+1)}) \leq g_{i+1}(\Omega_{v_{i+1}}) \max \left\{ d_2(H^{(i+1)}), 1 \right\},$$

where g_{i+1} is a certain computable function. We prove that the statement holds for $H^{(i)}$.

Suppose that the carrier base μ_i is quadratic and let $\nu_{i,1}, \ldots, \nu_{i,n_i}$ be the transfer bases of Ω_{v_i}. Then

$$|H^{(i)}(\mu_i)| - |H^{(i+1)}(\mu_i)| \leq \sum_{n=1}^{n_i} |H^{(i+1)}(\nu_{i,n})|,$$

where $\nu_{i,n}$, $n = 1, \ldots, n_i$ are bases of $\Omega_{v_{i+1}}$. Thus, by induction hypothesis, we get

$$\sum_{n=1}^{n_i} |H^{(i+1)}(\nu_{i,n})| \leq n_i g_{i+1}(\Omega_{v_{i+1}}) \max \left\{ d_2(H^{(i+1)}), 1 \right\}.$$

Notice that the sets of quadratic-coefficient bases of Ω_{v_i} and $\Omega_{v_{i+1}}$ coincide, thus $d_2(H^{(i+1)}) = d_2(H^{(i)})$. Therefore,

$$d_1(H^{(i)}) = \left(|H^{(i)}(\mu_i)| - |H^{(i+1)}(\mu_i)| \right) + d_1(H^{(i+1)}) \leq$$
$$\leq \left(|H^{(i)}(\mu_i)| - |H^{(i+1)}(\mu_i)| \right) + g_{i+1}(\Omega_{v_{i+1}}) \cdot \max \left\{ d_2(H^{(i+1)}), 1 \right\} \leq$$
$$\leq (n_i + 1) \cdot g_{i+1}(\Omega_{v_{i+1}}) \cdot \max \left\{ d_2(H^{(i+1)}), 1 \right\} = g_i(\Omega_{v_i}) \cdot \max \left\{ d_2(H^{(i)}), 1 \right\}.$$

Suppose now that the carrier base μ_i is a quadratic-coefficient base and let $\nu_{i,1}, \ldots, \nu_{i,n_i}$ be the transfer bases of the generalised equation Ω_{v_i}. Since the duals of the transferred quadratic bases become quadratic-coefficient and since $|H^{(i)}(\nu_{i,n})| \leq |H^{(i)}(\mu_i)|$, $n = 1, \ldots, n_i$, we get that

$$d_2(H^{(i+1)}) \leq L + |H^{(i+1)}(\mu_i)| + \sum_{n=1}^{n_i} |H^{(i)}(\nu_{n_i,n})| \leq$$
$$\leq L + (n_i + 1)|H^{(i)}(\mu_i)| \leq (n_i + 1) \left(L + |H^{(i)}(\mu_i)| \right) = (n_i + 1) \cdot d_2(H^{(i)}),$$

where $L = \sum_\lambda |H^{(i)}(\lambda)|$ and the sum is taken over all bases that are quadratic-coefficient in both Ω_{v_i} and $\Omega_{v_{i+1}}$. Therefore,

$$d_1(H^{(i)}) = \left(|H^{(i)}(\mu_i)| - |H^{(i+1)}(\mu_i)|\right) + d_1(H^{(i+1)}) \leq$$
$$\leq |H^{(i)}(\mu_i)| + g_{i+1}(\Omega_{v_{i+1}}) \cdot \max\left\{d_2(H^{(i+1)}), 1\right\} \leq$$
$$\leq |H^{(i)}(\mu_i)| + (n_i + 1) \cdot g_{i+1}(\Omega_{v_{i+1}}) \cdot \max\left\{d_2(H^{(i)}), 1\right\} \leq$$
$$\leq (n_i + 2) \cdot g_{i+1}(\Omega_{v_{i+1}}) \cdot \max\left\{d_2(H^{(i)}), 1\right\} = g_i(\Omega_{v_i}) \cdot \max\left\{d_2(H^{(i)}), 1\right\}.$$

The statement of the lemma follows. □

Recall, that by $\widetilde{\Omega}$ we denote the generalised equation obtained from Ω applying D 3. Consider the section of $\widetilde{\Omega}_{v_1}$ of the form $[1, \alpha(w)]$. The section $[1, \alpha(w)]$ lies in the quadratic part of $\widetilde{\Omega}_{v_1}$. Let B' be the set of quadratic bases that belong to $[1, \alpha(w)]$ and let $\mathfrak{V}'(\Omega_{v_1})$ be the group of automorphisms of $\mathbb{G}_{R(\Omega^*)}$ that are invariant with respect to the non-quadratic part of $\widetilde{\Omega}_{v_1}$ and act identically on all the bases which do not belong to B'. By definition, $\mathfrak{V}'(\Omega_{v_1}) \leq \mathfrak{V}(\Omega_{v_1})$. Thus the solution $H^{(1)}$ minimal with respect to $\mathfrak{V}(\Omega_{v_1})$ is also minimal with respect to $\mathfrak{V}'(\Omega_{v_1})$. By Lemma 7.19 we have

(7.9) $$d_1(H^{(1)}) \leq f_1(\Omega_{v_1}) \max\left\{d_2(H^{(1)}), 1\right\}.$$

Recall that (see Definition 4.8)

(7.10) $$d_{A\Sigma}(H) = \sum_{i=1}^{\alpha(w)-1} |H_i|, \quad \psi_{A\Sigma}(H) = \sum_{\mu \in \omega_1} |H(\mu)| - 2d_{A\Sigma}(H).$$

Our next goal is, using inequality (7.9), to give an upper bound of the length of the interval $d_{A\Sigma}(H^{(1)})$ in terms of the excess $\psi_{A\Sigma}$ and the function $f_1(\Omega_{v_1})$. More precisely, we have the following lemma.

LEMMA 7.20. *In the above notation, the following inequality holds*

(7.11) $$d_{A\Sigma}(H^{(1)}) \leq \max\left\{\psi_{A\Sigma}(H^{(1)})(2nf_1(\Omega_{v_1}) + 1), f_1(\Omega_{v_1})\right\}.$$

PROOF. Denote by $\gamma_i(w)$ the number of bases $\mu \in \omega_1$ containing h_i. Then

(7.12) $$\sum_{\mu \in \omega_1} |H^{(1)}(\mu)| = \sum_{i=1}^{\rho} |H_i^{(1)}| \gamma_i(w),$$

where $\rho = \rho_{\Omega_{v_1}}$. Let $I = \{i \mid 1 \leq i \leq \alpha(w) - 1 \text{ and } \gamma_i = 2\}$, and $J = \{i \mid 1 \leq i \leq \alpha(w) - 1 \text{ and } \gamma_i > 2\}$. By (4.3) we have:

(7.13) $$d_{A\Sigma}(H^{(1)}) = \sum_{i \in I} |H_i^{(1)}| + \sum_{i \in J} |H_i^{(1)}| = d_1(H^{(1)}) + \sum_{i \in J} |H_i^{(1)}|.$$

Let $\lambda, \Delta(\lambda)$ be a pair of variable quadratic-coefficient bases of the generalised equation $\widetilde{\Omega}_{v_1}$, where λ belongs to the non-quadratic part of $\widetilde{\Omega}_{v_1}$. When we apply D 3 to Ω_{v_1} thereby obtaining $\widetilde{\Omega}_{v_1}$, the pair $\lambda, \Delta(\lambda)$ is obtained from bases $\mu \in \omega_1$. There are two types of quadratic-coefficient bases.

Type 1: variable bases λ such that $\beta(\lambda) \leq \alpha(\omega)$. In this case, since λ belongs to the non-quadratic part of $\widetilde{\Omega}_{v_1}$, it is a product of items $\{h_i \mid i \in J\}$ and thus $|H(\lambda)| \leq \sum_{i \in J} |H_i^{(1)}|$. Thus the sum of the lengths of quadratic-coefficient bases of Type 1 and their duals is bounded above by $2n \sum_{i \in J} |H_i^{(1)}|$, where n is the number of bases in Ω.

Type 2: variable bases λ such that $\alpha(\lambda) \geq \alpha(\omega)$. The sum of the lengths of quadratic-coefficient bases of the second type is bounded above by
$$2 \cdot \sum_{i=\alpha(\omega)}^{\rho} |H_i^{(1)}| \gamma_i(\omega).$$

We have

(7.14) $$d_2(H^{(1)}) \leq 2n \sum_{i \in J} |H_i^{(1)}| + 2 \cdot \sum_{i=\alpha(\omega)}^{\rho} |H_i^{(1)}| \gamma_i(\omega).$$

Then from (7.10) and (7.12) it follows that

(7.15) $$\psi_{A\Sigma}(H_i^{(1)}) \geq \sum_{i \in J} |H_i^{(1)}| + \sum_{i=\alpha(\omega)}^{\rho} |H_i^{(1)}| \gamma_i(\omega).$$

From Equation (7.13), using inequalities (7.9), (7.14), (7.15) we get inequality (7.11). □

(III.2): **Minimal solutions H such that $\mathfrak{p}(H)$ contains a prohibited subpath of type 15 fail inequality** (7.15). Let the path $v_1 \to v_2 \to \ldots \to v_m$ corresponding to the sequence (7.8) be μ-reducing, that is $\mu_1 = \mu$ and, either there are no outgoing auxiliary edges from v_2 and μ occurs in the sequence $\mu_1, \ldots, \mu_{m-1}$ at least twice, or v_2 does have outgoing auxiliary edges $v_2 \to w_1, \ldots, v_2 \to w_n$ and the base μ occurs in the sequence $\mu_1, \ldots, \mu_{m-1}$ at least $\max\limits_{1 \leq i \leq n} \mathbf{s}(\Omega_{w_i})$ times.

Set $\delta_i = d_{A\Sigma}(H^{(i)}) - d_{A\Sigma}(H^{(i+1)})$. We give a lower bound for $\sum_{i=1}^{m-1} \delta_i$, i.e. we estimate by how much the length of a solution is reduced in a μ-reducing path.

We first prove that if $\mu_{i_1} = \mu_{i_2} = \mu$, $i_1 < i_2$ and $\mu_i \neq \mu$ for $i_1 < i < i_2$, then

(7.16) $$\sum_{i=i_1}^{i_2-1} \delta_i \geq |H^{(i_1+1)}[1, \alpha(\Delta(\mu_{i_1+1}))]|.$$

Indeed, if $i_2 = i_1 + 1$ then
$$\delta_{i_1} = |H^{(i_1)}[1, \alpha(\Delta(\mu))]| = |H^{(i_1+1)}[1, \alpha(\Delta(\mu))]|.$$

If $i_2 > i_1 + 1$, then $\mu_{i_1+1} \neq \mu$ and μ is a transfer base in the generalised equation $\Omega_{v_{i_1+1}}$ and thus

(7.17) $$\delta_{i_1+1} + |H^{(i_1+2)}[1, \alpha(\mu)]| = |H^{(i_1+1)}[1, \alpha(\Delta(\mu_{i_1+1}))]|.$$

Since μ is the carrier base of $\Omega_{v_{i_2}}$ we have

(7.18) $$\sum_{i=i_1+2}^{i_2-1} \delta_i \geq |H^{(i_1+2)}[1, \alpha(\mu)]|.$$

From (7.18) and (7.17) we get (7.16).

We want to show that every μ-reducing path reduces the length of a solution $H^{(1)}$ by at least $\frac{1}{10}|H^{(1)}(\mu)|$.

LEMMA 7.21. *Let $v_1 \to \cdots \to v_m$ be a μ-reducing path, then the following inequality holds*

$$\sum_{i=1}^{m-1} \delta_i \geq \frac{1}{10}|H^{(1)}(\mu)|. \tag{7.19}$$

PROOF. To prove the Lemma we consider the two cases from the definition of a μ-reducing path.

Suppose first that v_2 does not have any outgoing auxiliary edges, i.e. the bases μ_2 and $\Delta(\mu_2)$ do not intersect in the generalised equation Ω_{v_2}, then (7.16) implies that

$$\sum_{i=1}^{m-1} \delta_i \geq |H^{(2)}[1, \alpha(\Delta(\mu_2))]| \geq |H^{(2)}(\mu_2)| \geq |H^{(2)}(\mu)| = |H^{(1)}(\mu)| - \delta_1,$$

which, in turn, implies that

$$\sum_{i=1}^{m-1} \delta_i \geq \frac{1}{2}|H^{(1)}(\mu)|. \tag{7.20}$$

Suppose now that there are auxiliary edges $v_2 \to w_1, \ldots, v_2 \to w_n$. Let $H^{(2)}[1, \alpha(\Delta(\mu_2))] \doteq Q$, and P be a period such that $Q \doteq P^d$ for some $d \geq 1$, then $H^{(2)}(\mu_2)$ and $H^{(2)}(\mu)$ are initial subwords of the word $H^{(2)}[1, \beta(\Delta(\mu_2))]$, which, in turn, is an initial subword of P^∞.

By construction of the sequence (7.6), relation (7.7) fails for the vertex v_2, i. e. (in the notation of (7.7)):

$$H^{(2)}(\mu) \doteq P^r \cdot P_1, \ P \doteq P_1 \cdot P_2, \ r < \max_{1 \leq j \leq n} \mathbf{s}(\Omega_{w_j}). \tag{7.21}$$

Let $\mu_{i_1} = \mu_{i_2} = \mu$, $i_1 < i_2$ and $\mu_i \neq \mu$ for $i_1 < i < i_2$. If

$$|H^{(i_1+1)}(\mu_{i_1+1})| \geq 2|P| \tag{7.22}$$

since $H^{(i_1+1)}(\Delta(\mu_{i_1+1}))$ is a Q'-periodic subword (Q' is a cyclic permutation of P) of the Q'-periodic word $H^{(i_1+1)}[1, \beta(\Delta(\mu_{i_1+1}))]$ of length greater than $2|Q'| = 2|P|$, it follows by Lemma 1.2.9 in [Ad75], that

$$|H^{(i_1+1)}[1, \alpha(\Delta(\mu_{i_1+1}))]| \geq k|Q'|.$$

As $k \neq 0$ (μ_{i+1} and $\Delta(\mu_{i+1})$ do not form a pair of matched bases), so $|H^{(i_1+1)}[1, \alpha(\Delta(\mu_{i_1+1}))]| \geq |P|$. Together with (7.16) this gives that $\sum_{i=i_1}^{i_2-1} \delta_i \geq |P|$. The base μ occurs in the sequence $\mu_1, \ldots, \mu_{m-1}$ at least r times, so either (7.22) fails for some $i_1 \leq m-1$ or $\sum_{i=1}^{m-1} \delta_i \geq (r-3)|P|$.

If (7.22) fails, then from the inequality $|H^{(i+1)}(\mu_i)| \leq |H^{(i+1)}(\mu_{i+1})|$ and the definition of δ_i follows that

$$\sum_{i=1}^{i_1} \delta_i \geq |H^{(1)}(\mu)| - |H^{(i_1+1)}(\mu_{i_1+1})| \geq (r-2)|P|.$$

hence in both cases $\sum_{i=1}^{m-1} \delta_i \geq (r-3)|P|$.

Notice that for $i_1 = 1$, inequality (7.16) implies that $\sum_{i=1}^{m-1} \delta_i \geq |Q| \geq |P|$; so $\sum_{i=1}^{m-1} \delta_i \geq \max\{1, r-3\}|P|$. Together with (7.21) this implies that

$$\sum_{i=1}^{m-1} \delta_i \geq \frac{1}{5}|H^{(2)}(\mu)| = \frac{1}{5}(|H^{(1)}(\mu)| - \delta_1).$$

Finally we get that

$$\sum_{i=1}^{m-1} \delta_i \geq \frac{1}{10}|H^{(1)}(\mu)|.$$

From the above inequality and inequality (7.20), we see that for a μ-reducing path inequality (7.19) always holds. □

We thereby have shown that in any μ-reducing path the length of the solution is reduced by at least $\frac{1}{10}$ of the length of the carrier base μ.

Notice that by property (3) from Definition 7.14, we can assume that the carrier bases μ_i and their duals $\Delta(\mu_i)$ belong to the active part $A\Sigma = [1, \alpha(\omega)]$. Then, by Lemma 4.9 and by construction of the path (7.8), we have that

$$\psi_{A\Sigma}(H^{(1)}) = \cdots = \psi_{A\Sigma}(H^{(m)}) = \ldots$$

We denote this number by $\psi_{A\Sigma}$.

LEMMA 7.22. *Let $v_1 \to v_2 \to \cdots \to v_m$ be a prohibited path of type 15. By definition, $v_1 \to v_2 \to \cdots \to v_m$ can be presented in the form (7.5). Let κ be the length of the subpath $\mathfrak{p}_1\mathfrak{s}_1 \ldots \mathfrak{p}_l\mathfrak{s}_l$. Then there exists a carrier base $\mu \in \omega$ such that the following inequality holds*

(7.23) $$|H^{(\kappa)}(\mu)| \geq \frac{1}{2n}\psi_{A\Sigma},$$

where n is the number of bases in Ω.

PROOF. From the definition of $\psi_{A\Sigma}$, see (7.10), we get that $\sum_{\mu \in \omega_1} |H^{(m)}(\mu)| \geq \psi_{A\Sigma}$, hence the inequality $|H^{(m)}(\mu)| \geq \frac{1}{2n}\psi_{A\Sigma}$ holds for at least one base $\mu \in \omega_1$. Since $H^{(m)}(\mu) \doteq \left(H^{(m)}(\Delta(\mu))\right)^{\pm 1}$, we may assume that $\mu \in \omega \cup \tilde{\omega}$.

If $\mu \in \omega$, then inequality (7.23) trivially holds.

If $\mu \in \tilde{\omega}$, then by the third condition in the definition of a prohibited path of type 15 (see Definition 7.14) there exists $\kappa \leq i \leq m$ such that μ is a transfer base of Ω_{v_i}. Hence,

$$|H^{(\kappa)}(\mu_i)| \geq |H^{(i)}(\mu_i)| \geq |H^{(i)}(\mu)| \geq |H^{(m)}(\mu)| \geq \frac{1}{2n}\psi_{A\Sigma}.$$

□

Finally, from conditions (1) and (2) in the definition of a prohibited path of type 15, from the inequality $|H^{(i)}(\mu)| \geq |H^{(\kappa)}(\mu)|$, $1 \leq i \leq \kappa$, and from (7.19) and (7.23), it follows that

$$(7.24) \qquad \sum_{i=1}^{\kappa-1} \delta_i \geq \max\left\{\frac{1}{20n}\psi_{A\Sigma}, 1\right\} \cdot (40n^2 f_1 + 20n + 1).$$

By Equation (4.5), the sum in the left part of the inequality (7.24) equals $d_{A\Sigma}(H^{(1)}) - d_{A\Sigma}(H^{(\kappa)})$, hence

$$(7.25) \qquad d_{A\Sigma}(H^{(1)}) \geq \max\left\{\frac{1}{20n}\psi_{A\Sigma}, 1\right\} \cdot (40n^2 f_1 + 20n + 1),$$

which contradicts (7.11).

Therefore, the assumption that there are prohibited subpaths (7.8) of type 15 in the path (7.6) led to a contradiction. Hence, the path (7.6) does not contain prohibited subpaths. This implies that $v_i \in T_0(\Omega)$ for all $(\Omega_{v_i}, H^{(i)})$ in (7.6).

In particular, we have shown that final leaves w of the tree $T(\Omega)$ are, in fact, leaves of the tree $T_0(\Omega)$. Naturally, we call such leaves the *final leaves of $T_0(\Omega)$*.

(C): **The pair $(\Omega_w, H^{[w]})$, where $\mathrm{tp}(w) = 2$, satisfies the properties required in Proposition 7.18.** For all i, either $v_i = v_{i+1}$ and $|H^{[i+1]}| < |H^{[i]}|$, or $v_i \to v_{i+1}$ is an edge of a finite tree $T_0(\Omega)$. Hence the sequence (7.6) is finite. Let $(\Omega_w, H^{[w]})$ be its final term. We show that $(\Omega_w, H^{[w]})$ satisfies the properties required in the proposition.

Property (1) follows directly from the construction of $H^{[w]}$.

We now prove that property (2) holds. Let $\mathrm{tp}(w) = 2$ and suppose that Ω_w has non-constant non-active sections. It follows from the construction of (7.6) that if $[j, k]$ is an active section of $\Omega_{v_{i-1}}$ and is a non-active section of Ω_{v_i} then $H^{[i]}[j,k] \doteq H^{[i+1]}[j,k] \doteq \ldots \doteq H^{[w]}[j,k]$. Therefore, (7.7) and the definition of $\mathrm{s}(\Omega_v)$ imply that the word $h_1 \ldots h_{\rho_w}$ can be subdivided into subwords $h[1, i_1], \ldots, h[i_{l'-1}, i_{\rho_w}]$, such that for any l either $H^{[w]}[i_l, i_{l+1}]$ has length 1, or the word $h[i_l, i_{l+1}]$ does not appear in basic and coefficient equations, or

$$(7.26) \qquad H^{[w]}[i_l, i_{l+1}] \doteq P_l^r \cdot P_l'; \quad P_l \doteq P_l' P_l''; \quad r \geq \rho_w \max_{\langle \mathcal{P}, R\rangle} \{f_0(\Omega_w, \mathcal{P}, R)\},$$

where P_l is a period, and the maximum is taken over all regular periodic structures of $\widetilde{\Omega}_w$. Therefore, if we choose P_l of maximal length, then $\widetilde{\Omega}_w$ is singular or strongly singular with respect to the periodic structure $\mathcal{P}(H^{[w]}, P_l)$. Indeed, suppose that it is regular with respect to this periodic structure. Then, as in $H^{[w]}[i_l, i_{l+1}]$ one has $i_{l+1} - i_l \leq \rho_w$, so (7.26) implies that there exists h_k such that $|H_k^{[w]}| \geq f_0(\Omega_w, \mathcal{P}, R)$. By Lemma 6.20, this contradicts the minimality of the solution $H^{[w]}$.

This finishes the proof of Proposition 7.18.

CHAPTER 8

From the coordinate group $\mathbb{G}_{R(\Omega^*)}$ to proper quotients: the decomposition tree T_{dec} and the extension tree T_{ext}

8.1. The decomposition tree $T_{\text{dec}}(\Omega)$

We proved in the previous chapter that for every solution H of a generalised equation Ω, the path $\mathfrak{p}(H)$ associated to the solution H ends in a final leaf v of the tree $T_0(\Omega)$, $\text{tp}(v) = 1, 2$. Furthermore, if $\text{tp}(v) = 2$ and the generalised equation Ω_v contains non-constant non-active sections, then the generalised equation Ω_v is singular or strongly singular with respect to the periodic structure $\mathcal{P}(H^{(v)}, P)$, see Proposition 7.18.

The essence of the decomposition tree $T_{\text{dec}}(\Omega)$ is that to every solution H of Ω one can associate the path $\mathfrak{p}(H)$ in $T_{\text{dec}}(\Omega)$ such that either all sections of the generalised equation Ω_u corresponding to the leaf u of $T_{\text{dec}}(\Omega)$ are non-active constant sections or the coordinate group of Ω_u is a proper quotient of $\mathbb{G}_{R(\Omega^*)}$.

We summarise the results of this section in the proposition below.

PROPOSITION 8.1. *For a (constrained) generalised equation $\Omega = \Omega_{v_0}$, one can effectively construct a finite oriented rooted at v_0 tree T_{dec}, $T_{\text{dec}} = T_{\text{dec}}(\Omega_{v_0})$, such that:*

(1) *The tree $T_0(\Omega)$ is a subtree of the tree T_{dec}.*
(2) *To every vertex v of T_{dec} we assign a recursive group of automorphisms $A(\Omega_v)$.*
(3) *For any solution H of a generalised equation Ω there exists a leaf u of the tree T_{dec}, $\text{tp}(u) = 1, 2$, and a solution $H^{[u]}$ of the generalised equation Ω_u such that*
   - $\pi_H = \sigma_0 \pi(v_0, v_1) \sigma_1 \ldots \pi(v_{n-1}, u) \sigma_n \pi_{H^{[u]}}$, *where $\sigma_i \in A(\Omega_{v_i})$;*
   - *if $\text{tp}(u) = 2$, then all non-active sections of Ω_u are constant sections.*

To obtain $T_{\text{dec}}(\Omega)$ we add some edges labelled by proper epimorphisms (to be described below) to the final leaves of type 2 of $T_0(\Omega)$ so that the corresponding generalised equations contain non-constant non-active sections.

The idea behind the construction of this tree is the following. Lemmas 6.18 and 6.17 state that given a generalised equation Ω singular or strongly singular with respect to a periodic structure $\langle \mathcal{P}, R \rangle$, there exist finitely many proper quotients of the coordinate group $\mathbb{G}_{R(\Omega^*)}$ such that for every P-periodic solution H of the generalised equation Ω such that $\mathcal{P}(H, P) = \langle \mathcal{P}, R \rangle$, an $\mathfrak{A}(\Omega)$-automorphic image H^+ of H is a solution of a proper equation. In other words, the $\mathbb{G}$-homomorphism

$\pi_{H^+} : \mathbb{G}_{R(\Omega^*)} \to \mathbb{G}$ factors through one of the finitely many proper quotients of $\mathbb{G}_{R(\Omega^*)}$.

Recall that given a coordinate group of a system of equations, the homomorphisms π from this coordinate group to the coordinate groups of constrained generalised equations determined by partition tables (see Equation (3.2) and discussion in Section 3.2.3), are, in general, just homomorphisms (neither injective nor surjective). Our goal here is to prove that in fact, the solution H^+ is a solution of one of the finitely many proper generalised equations such that the homomorphism from the coordinate group of the system of equations to the coordinate group of the generalised equation is an epimorphism.

Let v be a final leaf of type 2 of $T_0(\Omega)$ such that Ω_v contains non-constant non-active sections. For every periodic structure $\langle \mathcal{P}, R \rangle$ on Ω_v such that Ω_v is either singular or strongly singular with respect to $\langle \mathcal{P}, R \rangle$, we now describe the vertices that we introduce in the tree T_{dec} (these vertices will be leaves of T_{dec}), the generalised equations associated to these new vertices and the epimorphisms corresponding to the edges that join the leaf v with these vertices. We then prove that for every $\mathcal{P}$-periodic solution H of the generalised equation Ω_v such that Ω_v is singular or strongly singular with respect to $\mathcal{P}(H, P)$, there exists an $\mathfrak{A}(\Omega)$-automorphic image H^+ of H so that H^+ is a solution of one of the generalised equations associated to the new vertices.

We first consider the case when Ω_v is strongly singular of type (b) with respect to the periodic structure $\langle \mathcal{P}, R \rangle$ (see Definition 6.15).

In Lemma 6.17, we proved that every $\mathcal{P}$-periodic solution H of a generalised equation Ω strongly singular (of type (b)) with respect to a periodic structure $\langle \mathcal{P}, R \rangle$ such that $\mathcal{P}(H, P) = \langle \mathcal{P}, R \rangle$, is a solution of a system of equations $\bar{\Omega}$ obtained from Ω by adding new equations. We now explicitly construct a single generalised equation $\Omega_v(\mathcal{P}, R, \{g_i\})$ so that every $\mathcal{P}$-periodic solution H of a generalised equation Ω strongly singular (of type (b)) with respect to a periodic structure $\langle \mathcal{P}, R \rangle$ such that $\mathcal{P}(H, P) = \langle \mathcal{P}, R \rangle$, is a solution of $\Omega_v(\mathcal{P}, R, \{g_i\})$.

Let $g_q \in \{g_i\}$ be an element of the family $\{g_i\}$ constructed in Lemma 6.17. As shown in the proof of Lemma 6.17, the element g_q is a commutator of the form $[\varphi_q(h_{i,q}), h_{j,q}]$, where φ_q is a generator of the automorphism group $\mathfrak{A}(\Upsilon_v)$, see parts (2), (3) of Lemma 6.14 and Definition 6.16. Let $\varphi_q(h_{i,q}) = h_{i_1,q} \cdots h_{i_{k_q},q}$. Define the generalised equation $\Omega_v(\mathcal{P}, R, \{g_i\}) = \langle \Upsilon_v(\mathcal{P}, R, \{g_i\}), \Re_{\Upsilon_v(\mathcal{P},R,\{g_i\})} \rangle$ as follows. Set $\Upsilon_v(\mathcal{P}, R, \{g_i\}) = \Upsilon_v$ and define $\Re_{\Upsilon_v(\mathcal{P},R,\{g_i\})}$ to be the minimal subset of $h \times h$ that contains $\Re_{\Upsilon_v}$, $\{(h_{i_l,q}, h_{j,q}) \mid l = 1, \ldots, k_q\}$ for all q, is symmetric and satisfies the condition $(\star)$ from Definition 3.4.

The natural homomorphism $\pi_{v,\{g_i\}} : \mathbb{G}_{R(\Omega_v^*)} \to \mathbb{G}_{R(\Omega_v(\mathcal{P},R,\{g_i\})^*)}$ is surjective. Moreover, by construction, $\pi_{v,\{g_i\}}(h(g_q)) = 1$ for all $g_q \in \{g_i\}$, and therefore $\pi_{v,\{g_i\}}$ is a proper epimorphism.

We introduce a new vertex w and associate the generalised equation $\Omega_w = \Omega_v(\mathcal{P}, R, \{g_i\})$ to it. The vertex w is a leaf of the tree T_{dec} and is joined to the vertex v by an edge $v \to w$ labelled by the epimorphism $\pi(v, w) = \pi_{v,\{g_i\}}$.

We now show that every $\mathcal{P}$-periodic solution H of Ω_v such that Ω_v is strongly singular of type (b) with respect to $\mathcal{P}(H, P)$, is a solution of the generalised equation Ω_w. Obviously, H is a solution of Υ_w. We are left to prove that if $\Re_{\Upsilon_w}(h_i, h_j)$, then $H_i \leftrightarrows H_j$.

8.1. THE DECOMPOSITION TREE $T_{\text{dec}}(\Omega)$

In the proof of Lemma 6.17, on the one hand we have shown that $H_{i,q} \leftrightarrows H_{j,q}$ and that $\text{alph}(H_{i,q}) = \text{alph}(P)$, thus $\text{alph}(P) \leftrightarrows H_{j,q}$. On the other hand, we have shown that for every generator φ_q of the automorphism group $\mathfrak{A}(\Omega_v)$, the image $\varphi_q(h_{i,q})$ is a word $h_{i_1,q} \cdots h_{i_k,q}$ in variables $\{h_{i_l,q} \mid h_{i_l,q} \in \sigma, \sigma \in \mathcal{P}, l = 1, \ldots, k\}$. Since for every solution H on has $\text{alph}(H_{i_l,q}) \subseteq \text{alph}(P)$ and $\text{alph}(P) \leftrightarrows H_{j,q}$, we get that $H_{i_l,q} \leftrightarrows H_{j,q}$ and the statement follows.

Let us consider the case when Ω_v is strongly singular of type (a) with respect to the periodic structure $\langle \mathcal{P}, R \rangle$ (see Definition 6.15).

In Lemma 6.17, we proved that every $\mathcal{P}$-periodic solution H of a generalised equation Ω strongly singular (of type (a)) with respect to a periodic structure $\langle \mathcal{P}, R \rangle$ such that $\mathcal{P}(H, \mathcal{P}) = \langle \mathcal{P}, R \rangle$, is a solution of the proper system of equations $\{\Omega_v \cup \{g_i\}\}^*$. Notice that, a priori, H is a solution of the system of equations $\{\Omega_v \cup \{g_i\}\}^*$ over $\mathbb{G}$. Our goal is to construct a generalised equation $\Omega_v(\mathcal{P}, R, \{g_i\}, \mathcal{T})$ in such a way that H is a solution of the generalised equation $\Omega_v(\mathcal{P}, R, \{g_i\}, \mathcal{T})$ and the homomorphism from $\mathbb{G}_{R(\Omega_v^*)}$ to $\mathbb{G}_{R(\Omega_v(\mathcal{P},R,\{g_i\},\mathcal{T})^*)}$ is a proper epimorphism.

Below we use the notation of Section 3.2. For the system of equations $\{g_i = 1\}$, where the elements g_i are defined in Lemma 6.17, consider the subset $\mathcal{PT}'$ of the set $\mathcal{PT}$ of all $\mathbb{G}$-partition tables of the system $\{g_i\}$, of the form $(V, \mathbb{G} * F(z_1, \ldots, z_p))$ that satisfy the following condition

(8.1) $$z_1, \ldots, z_p \in \langle V_{i,j} \rangle.$$

For any generalised equation $\Upsilon_{\mathcal{T}}$, $\mathcal{T} \in \mathcal{PT}'$, the homomorphism $\pi_{\Upsilon_{\mathcal{T}}} : \mathbb{G}_{R(\{g_i\})} \to \mathbb{G}_{R(\Upsilon_{\mathcal{T}}^*)}$ induced by the map $h_i \mapsto P_{h_i}(h', \mathcal{A})$ (see Section 3.2.3 for definition) is surjective. Consider the set of generalised equations $\{\Upsilon_{\mathcal{T}} \mid \mathcal{T} \in \mathcal{PT}'\}$ over the free monoid $\mathbb{F}$. Note that in general, this set of generalised equations should be considered over $\mathbb{T}$. However, we will show that for $\mathcal{P}$-periodic solutions H it suffices to consider the set of generalised equations $\{\Upsilon_{\mathcal{T}} \mid \mathcal{T} \in \mathcal{PT}'\}$ over $\mathbb{F}$.

Construct the generalised equation

$$\Omega_v(\mathcal{P}, R, \{g_i\}, \mathcal{T}) = \langle \Upsilon_v(\mathcal{P}, R, \{g_i\}, \mathcal{T}), \Re_{\Upsilon_v(\mathcal{P}, R, \{g_i\}, \mathcal{T})} \rangle$$

as follows. The set of items $\tilde{h} = h \cup h'$ of $\Omega_v(\mathcal{P}, R, \{g_i\}, \mathcal{T})$ is the disjoint union of the items of Υ_v and $\Upsilon_{\mathcal{T}}$; the set of coefficient equations of $\Omega_v(\mathcal{P}, R, \{g_i\}, \mathcal{T})$ is the disjoint union of the coefficient equations of Υ_v and $\Upsilon_{\mathcal{T}}$. The set of basic equations of $\Omega_v(\mathcal{P}, R, \{g_i\}, \mathcal{T})$ consists of: the basic equations of Υ_v, the basic equations of $\Upsilon_{\mathcal{T}}$, and basic equations of the form $h_k = P_{h_k}(h', \mathcal{A})$, where in the left hand side of this equation h_k is treated as a variable of Υ_v and $P_{h_k}(h', \mathcal{A})$ is a label of a section of $\Upsilon_{\mathcal{T}}$.

The natural homomorphism $\pi'_{v,\{g_i\},\mathcal{T}} : \mathbb{G}_{R(\Upsilon_v^*)} \to \mathbb{G}_{R(\Upsilon_v(\mathcal{P},R,\{g_i\},\mathcal{T})^*)}$, induced by the map

$$\varpi : h_i \mapsto P_{h_i}(h', \mathcal{A}),$$

is surjective, since $\pi_{\Upsilon_{\mathcal{T}}}$ is.

The set of relations $\Re_{\Upsilon_v(\mathcal{P},R,\{g_i\},\mathcal{T})} \subseteq \tilde{h} \times \tilde{h}$ is defined as follows. Let $\varpi(h_i) = h'_{i_1} \cdots h'_{i_k}$ and $\varpi(h_j) = h'_{j_1} \cdots h'_{j_l}$. If $\Re_{\Upsilon_v}(h_i, h_j)$, then we set $\Re_{\Upsilon_v(\mathcal{P},R,\{g_i\},\mathcal{T})}(h'_{i_m}, h'_{j_n})$ for all $m = 1, \ldots, k$ and $n = 1, \ldots, l$. The set $\Re_{\Upsilon_v(\mathcal{P},R,\{g_i\},\mathcal{T})}$ is defined as the minimal subset of $\tilde{h} \times \tilde{h}$ that contains the above defined set, the set $\Re_{\Upsilon_v}$, is symmetric and satisfies the condition $(\star)$ from Definition

3.4. Note that by condition $(\star)$, the set $\Re_{\Upsilon_v(\mathcal{P},R,\{g_i\},\mathcal{T})}$ is independent of the choice of the map ϖ.

The natural homomorphism $\pi_{v,\{g_i\},\mathcal{T}} : \mathbb{G}_{R(\Omega_v^*)} \to \mathbb{G}_{R(\Omega_v(\mathcal{P},R,\{g_i\},\mathcal{T})^*)}$ is surjective, since $\pi'_{v,\{g_i\},\mathcal{T}}$ is. Moreover, by construction, $\pi_{v,\{g_i\},\mathcal{T}}(h(g_q)) = 1$ for all $g_q \in \{g_i\}$, and therefore $\pi_{v,\{g_i\},\mathcal{T}}$ is a proper epimorphism.

For every partition table $\mathcal{T} \in \mathcal{PT}'$ as above, we introduce a new vertex $w_{\mathcal{T}}$ and associate the generalised equation $\Omega_{w_{\mathcal{T}}} = \Omega_v(\mathcal{P}, R, \{g_i\}, \mathcal{T})$ to this vertex. The vertex $w_{\mathcal{T}}$ is a leaf of the tree T_{dec} and is joined to v by an edge $v \to w_{\mathcal{T}}$ labelled by the epimorphism $\pi_{v,\{g_i\},\mathcal{T}}$.

We now show that every P-periodic solution H of Ω_v such that Ω_v is strongly singular of type (a) with respect to $\mathcal{P}(H,P)$ factors through one of the solutions of generalised equations $\Omega_v(\mathcal{P}, R, \{g_i\}, \mathcal{T})$ for some choice of $\mathcal{T}$. By Lemma 6.8, the solution H is a solution of the system of equations $\{g_q = [h(\mathfrak{c}_{1,q}), h(\mathfrak{c}_{2,q})] = 1 \mid g_q \in \{g_i\}\}$. It follows that if we write the system $\{[h(\mathfrak{c}_{1,q}), h(\mathfrak{c}_{2,q})] = 1 \mid g_q \in \{g_i\}\}$ in the form (3.1), and if $r_1 r_2 \cdots r_l = 1$ is an equation of this system and $H_{j_1}, \ldots, H_{j_l}$ are the respective components of H, then the word $H_{j_1} \cdots H_{j_l}$ is trivial in the free group $F(\mathcal{A})$, and in the product of two consecutive subwords H_{j_r} and $H_{j_{r+1}}$ either there is no cancellation, or one of these words cancels completely, that is $H_{j_r} \doteq W \cdot (H_{j_{r+1}})^{-1}$ or, vice versa $H_{j_{r+1}} \doteq (H_{j_r})^{-1} \cdot W$. This shows that the $\mathbb{G}$-partition table of the system $\{[h(\mathfrak{c}_{1,q}), h(\mathfrak{c}_{2,q})] = 1 \mid g_q \in \{g_i\}\}$ for which the solution H factors through satisfies condition (7.2).

Finally, we consider the case when Ω_v is singular with respect to the periodic structure $\langle \mathcal{P}, R \rangle$ (see Definition 6.15).

In Lemma 6.18, we proved that for every P-periodic solution H of a generalised equation Ω singular with respect to a periodic structure $\langle \mathcal{P}, R \rangle$ such that $\mathcal{P}(H,P) = \langle \mathcal{P}, R \rangle$ there exists an $\mathfrak{A}(\Omega)$-automorphic image H^+ of H that is a solution of a proper equation. Though H^+ is a solution of Ω^*, it may be not a solution of Ω, see the example given in Section 6.4. We construct the generalised equation $\Omega_v(\mathcal{P}, R, \mathfrak{c}, \mathcal{T})$ in such a way that H^+ induces a solution of $\Omega_v(\mathcal{P}, R, \mathfrak{c}, \mathcal{T})$ and the homomorphism from $\mathbb{G}_{R(\Omega_v^*)}$ to $\mathbb{G}_{R(\Omega_v(\mathcal{P},R,\mathfrak{c},\mathcal{T})^*)}$ is a proper epimorphism.

More precisely, consider a triple $(\langle \mathcal{P}, R \rangle, \mathfrak{c}, \mathcal{T})$, where

- Ω_v is singular with respect to the periodic structure $\langle \mathcal{P}, R \rangle$,
- $\mathfrak{c}$ is a cycle in Γ, $\mathfrak{c} \in \{\mathfrak{c}_1, \ldots, \mathfrak{c}_r\}$, where $\{\mathfrak{c}_1, \ldots, \mathfrak{c}_r\}$ is the set of cycles from Lemma 6.18,
- and $\mathcal{T}$ is a partition table from the set $\mathcal{PT}'$ defined below.

For every such triple, we construct a generalised equation $\Omega_v(\mathcal{P}, R, \mathfrak{c}, \mathcal{T})$ and a proper epimorphism $\pi_{v,\mathfrak{c},\mathcal{T}} : \mathbb{G}_{R(\Omega_v^*)} \to \mathbb{G}_{R(\Omega_v(\mathcal{P},R,\mathfrak{c},\mathcal{T})^*)}$, put an edge $v \to u$, $\Omega_u = \Omega_v(\mathcal{P}, R, \mathfrak{c}, \mathcal{T})$ and show that every solution H^+ of Ω_v is a solution of one of the generalised equations $\Omega_v(\mathcal{P}, R, \mathfrak{c}, \mathcal{T})$ for some $\mathfrak{c}$ and $\mathcal{T}$. The vertex u is a leaf of $T_{\text{dec}}(\Omega)$.

The way we proceed is the following. By part (3) of Lemma 6.18, we know that H^+ is a solution of the generalised equation $\{h(\mu) = h(\Delta(\mu)) \mid \mu \notin \mathcal{P}\}$. This fact pilots the construction of the generalised equation $\hat{\Upsilon}_v$ below. On the other hand, we consider the system of equations $\mathbf{S}$ over $\mathbb{G}$ corresponding to the bases from $\mathcal{P}$ and the cycle $\mathfrak{c}$ (see below). Since the solution H^+ satisfies statement (3) of Lemma 6.18, we can prove that the $\mathbb{G}$-partition table $\mathcal{T}$ associated to H^+ satisfies property (7.2). For the partition tables $\mathcal{T}$ that satisfy condition (7.2) and the corresponding generalised equations $\Upsilon_{\mathcal{T}}$, the homomorphism $\pi_{\Upsilon_{\mathcal{T}}} : \mathbb{G}_{R(\mathbf{S})} \to \mathbb{G}_{R(\Upsilon_{\mathcal{T}}^*)}$ is surjective. We construct the generalised equation $\Omega_v(\mathcal{P}, R, \mathfrak{c}, \mathcal{T})$ from $\Upsilon_{\mathcal{T}}$ and $\check{\Upsilon}_v$ by identifying the items they have in common. We refer the reader to Section 8.2 for an example of the constructions given below.

We now formalise the construction of $\mathbb{G}_{R(\Omega_v(\mathcal{P}, R, \mathfrak{c}, \mathcal{T})^*)}$ and $\pi_{v,\mathfrak{c},\mathcal{T}}$.

Let $\mathfrak{c}$ be as above and suppose that $h(\mathfrak{c}) = h_{i_1} \cdots h_{i_k}$. Consider the system of equations $\mathbf{S} \subset \mathbb{G}[h]$ over $\mathbb{G}$:

$$\mathbf{S} = \{h(\mathfrak{c}_\mu) = 1, h(\mathfrak{c}) = h_{i_1} \cdots h_{i_k} = 1 \mid \mu \in \mathcal{P}\}.$$

Below we use the notation of Section 3.2. For the system of equations $\mathbf{S}$ consider the subset $\mathcal{PT}'$ of the set $\mathcal{PT}(\mathbf{S})$ of all $\mathbb{G}$-partition tables of $\mathbf{S}$ of the form $(V, \mathbb{G} * F(z_1, \ldots, z_p))$ that satisfy condition (7.2). Consider the set of generalised equations $\{\Upsilon_{\mathcal{T}} \mid \mathcal{T} \in \mathcal{PT}'\}$ over the free monoid $\mathbb{F}$. Note that, in general, this set of generalised equations should be considered over $\mathbb{T}$. However, we will show that for the minimal solution H^+ it suffices to consider the set of generalised equations $\{\Upsilon_{\mathcal{T}} \mid \mathcal{T} \in \mathcal{PT}'\}$ over $\mathbb{F}$. For any generalised equation $\Upsilon_{\mathcal{T}}$, $\mathcal{T} \in \mathcal{PT}'$, since the partition table $\mathcal{T}$ satisfies condition (7.2), it follows that the homomorphism $\pi_{\Upsilon_{\mathcal{T}}} : \mathbb{G}_{R(\mathbf{S})} \to \mathbb{G}_{R(\Upsilon_{\mathcal{T}}^*)}$ induced by the map $h_i \mapsto P_{h_i}(h', \mathcal{A})$ (see Section 3.2.3 for definition) is surjective.

Let the generalised equation $\tilde{\Upsilon}_v$ be obtained from $\check{\Upsilon}_v$ by removing all bases and items that belong to $\mathcal{P}$.

Construct the generalised equation

$$\Omega_v(\mathcal{P}, R, \mathfrak{c}, \mathcal{T}) = \langle \Upsilon_v(\mathcal{P}, R, \mathfrak{c}, \mathcal{T}), \Re_{\Upsilon_v(\mathcal{P}, R, \mathfrak{c}, \mathcal{T})} \rangle$$

as follows. The set of items $\tilde{h}$ of $\Omega_v(\mathcal{P}, R, \mathfrak{c}, \mathcal{T})$ is the disjoint union of the items of $\check{\Upsilon}_v$ and $\Upsilon_{\mathcal{T}}$; the set of coefficient equations of $\Omega_v(\mathcal{P}, R, \mathfrak{c}, \mathcal{T})$ is the disjoint union of the coefficient equations of $\check{\Upsilon}_v$ and $\Upsilon_{\mathcal{T}}$. The set of basic equations of $\Omega_v(\mathcal{P}, R, \mathfrak{c}, \mathcal{T})$ consists of: the basic equations of $\check{\Upsilon}_v$, the basic equation of $\Upsilon_{\mathcal{T}}$, and basic equations of the form $h_k = P_{h_k}(h', \mathcal{A})$, $h_k \notin \mathcal{P}$, where in the left hand side of this equation h_k is treated as a variable of $\check{\Upsilon}_v$ and $P_{h_k}(h', \mathcal{A})$, $h_k \notin \mathcal{P}$ is a label of a section of $\Upsilon_{\mathcal{T}}$.

The natural homomorphism $\pi'_{v,\mathfrak{c},\mathcal{T}} : \mathbb{G}_{R(\Upsilon_v^*)} \to \mathbb{G}_{R(\Upsilon_v(\mathcal{P}, R, \mathfrak{c}, \mathcal{T})^*)}$, induced by a map ϖ:

$$h_i \mapsto \begin{cases} P_{h_i}(h', \mathcal{A}), & \text{when } h_i \in \mathcal{P}, \\ h_i, & \text{otherwise;} \end{cases}$$

is surjective, since $\pi_{\Upsilon_{\mathcal{T}}}$ is.

The relation $\Re_{\Upsilon_v(\mathcal{P}, R, \mathfrak{c}, \mathcal{T})} \subseteq \tilde{h} \times \tilde{h}$ is defined as follows. Let $\varpi(h_i) = h'_{i_1} \cdots h'_{i_k}$ and $\varpi(h_j) = h'_{j_1} \cdots h'_{j_l}$. If $\Re_{\Upsilon_v}(h_i, h_j)$, then we set $\Re_{\Upsilon_v(\mathcal{P}, R, \mathfrak{c}, \mathcal{T})}(h'_{i_m}, h'_{j_n})$ for all $m = 1, \ldots, k$ and $n = 1, \ldots, l$. The set $\Re_{\Upsilon_v(\mathcal{P}, R, \mathfrak{c}, \mathcal{T})}$ is defined as the minimal subset of $\tilde{h} \times \tilde{h}$ that contains the above defined set, the set $\Re_{\tilde{\Upsilon}_v}$ (the restriction of the relation $\Re_{\Upsilon_v}$ onto the set of items of $\tilde{\Upsilon}_v$), is symmetric and satisfies the condition $(\star)$ from Definition 3.4. Note that, by condition $(\star)$, the set $\Re_{\Upsilon_v(\mathcal{P}, R, \mathfrak{c}, \mathcal{T})}$ is independent of the choice of the map ϖ.

The natural homomorphism $\pi_{v,\mathfrak{c},\mathcal{T}} : \mathbb{G}_{R(\Omega_v^*)} \to \mathbb{G}_{R(\Omega_v(\mathcal{P}, R, \mathfrak{c}, \mathcal{T})^*)}$ is surjective, since $\pi'_{v,\mathfrak{c},\mathcal{T}}$ is. Moreover, by construction, $\pi_{v,\mathfrak{c},\mathcal{T}}(h(\mathfrak{c})) = 1$, and therefore $\pi_{v,\mathfrak{c},\mathcal{T}}$ is a proper epimorphism.

We now show that for every P-periodic solution H of Ω_v such that Ω_v is singular with respect to the periodic structure $\mathcal{P}(H, P)$, there exists an $\mathfrak{A}(\Omega_v)$-automorphic image H^+ of H such that H^+ factors through one of the solutions of the generalised equations $\Omega_v(\mathcal{P}, R, \mathfrak{c}, \mathcal{T})$, for some choice of $\mathfrak{c}$ and $\mathcal{T}$. Indeed, let H^+ be the solution constructed in Lemma 6.18 and $\mathfrak{c}$ be one of the cycles from Lemma 6.18 for which $H^+(\mathfrak{c}) = 1$. The solution H^+ is a solution of the system **S**. All equations of **S** have the form $h(\mathfrak{c}') = 1$, where $\mathfrak{c}'$ is a cycle in the graph of the periodic structure $\mathcal{P}(H, P)$. From condition (3) of Lemma 6.18, it follows that, on the one hand, the word H_k^+, $1 \le k \le \rho$, is geodesic, and, on the other hand, that if $r_1 r_2 \cdots r_l = 1$ is an equation of the system **S** and $H_{j_1}^+, \ldots, H_{j_l}^+$ are the respective components of H^+, then the word $H_{j_1}^+ \cdots H_{j_l}^+$ is trivial in the free group $F(\mathcal{A})$. Furthermore, in the product of two consecutive subwords $H_{j_r}^+$ and $H_{j_{r+1}}^+$ either there is no cancellation, or one of these words cancels completely, that is either $H_{j_r}^+ \doteq W \cdot \left(H_{j_{r+1}}^+\right)^{-1}$ or $H_{j_{r+1}} \doteq (H_{j_r})^{-1} \cdot W$. Let $\mathcal{T}$ be the partition table corresponding to the cancellation scheme of the system **S**. The argument above shows that $\mathcal{T}$ satisfies condition (7.2).

Let $v_0 = v$, $\Omega_{v_1} = \Omega_v(\mathcal{P}, R, \mathfrak{c}, \mathcal{T})$.

The solution H^+ induces a solution $H_{\mathbf{S}}^+$ of the system **S**. By Lemma 3.21, there exists a solution $H_{\Upsilon_\mathcal{T}}^+$ of the generalised equation $\Upsilon_\mathcal{T}$ such that the following diagram

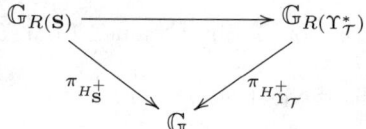

is commutative.

As H^+ satisfies condition (3) of Lemma 6.18, and the H_i's are geodesic, so the H_i^+'s are geodesic. In particular, $H_{\Upsilon_\mathcal{T}}^+$ is a solution of the generalised equation $\Upsilon_\mathcal{T}$ over the free monoid $\mathbb{F}$.

On the other hand, using again part (3) of Lemma 6.18, if we remove the components $\{H_k^+ \mid h_k \in \mathcal{P}\}$ of the solution H^+, we get a solution $\check{H}^+$ of the generalised equation $\check{\Upsilon}_v$.

Combining the solutions $\check{H}^+$ and $H_{\Upsilon_\mathcal{T}}^+$ we get a solution $H^{(v_1)}$ of the generalised equation $\Upsilon_v(\mathcal{P}, R, \mathfrak{c}, \mathcal{T})$. By part (3) of Lemma 6.18, the solution H^+ satisfies the commutation constraints from $\Re_{\Upsilon_v}$. Therefore, from the construction it follows that $H^{(v_1)}$ is a solution of $\Omega_v(\mathcal{P}, R, \mathfrak{c}, \mathcal{T})$ and

$$\pi_{H^+} = \pi(v_0, v_1) \pi_{H^{(v_1)}}.$$

We thereby have shown that for every P-periodic solution H of Ω_v such that Ω_v is singular with respect to the periodic structure $\mathcal{P}(H, P)$, there exists an $\mathfrak{A}(\Omega_v)$-automorphic image H^+ of H such that H^+ factors through one of the solutions of the generalised equations $\Omega_v(\mathcal{P}, R, \mathfrak{c}, \mathcal{T})$.

To the root vertex v_0 of the decomposition tree $T_{\text{dec}(\Omega)}$ we associate the group of automorphisms $\text{Aut}(\Omega)$ of the coordinate group $\mathbb{G}_{R(\Omega_{v_0})}$, see Definition 7.16. To the vertices v such that v is a leaf of $T_0(\Omega)$ but not of $T_{\text{dec}}(\Omega)$ (those vertices to which we added edges), we associate the group of automorphisms generated by the groups $\mathfrak{A}(\Omega_v)$, see Definition 6.16, corresponding to all periodic structures on Ω_v with respect to which Ω_v is singular or strongly singular. To all the other vertices of $T_{\text{dec}}(\Omega)$ we associate the trivial group of automorphisms. We denote the automorphism group associated to a vertex v of the tree $T_{\text{dec}}(\Omega)$ by $A(\Omega_v)$.

Naturally, we call leaves u such that the paths $\mathfrak{p}(H)$ associated to solutions H of Ω end in u, *final leaves of $T_{\text{dec}}(\Omega)$*.

8.2. Example

Let Ω be the generalised equation shown on Figure 12:
$$h_1h_2h_3h_4 = h_4h_5h_6h_7;\ h_1 = h_5;\ h_3 = h_7;\ h_2 = h_8;\ h_6 = h_8;\ h_8 = a.$$
Consider the periodic structure $\langle \mathcal{P}, R \rangle$ on Ω defined in Section 6.2. We have shown in the example given in Section 6.2 that this periodic structure is singular.

In the example given in Section 6.4, we considered the solution H:
$$\begin{array}{llll} H_1 = (bac)^2 b; & H_3 = (cba)^2 c; & H_5 = (bac)^2 b; & H_7 = (cba)^2 c; \\ H_2 = a; & H_4 = (bac)^6; & H_6 = a; & H_8 = a. \end{array}$$

In Section 6.4, for the solution H we constructed its automorphic image H^+ that satisfies conditions (2) and (3) of Lemma 6.18:
$$\begin{array}{llll} H_1^+ = (bac)^2 b; & H_3^+ = a^{-1}b^{-1}(bac)^{-2}; & H_5^+ = (bac)^2 b; & H_7^+ = a^{-1}b^{-1}(bac)^{-2}; \\ H_2^+ = a; & H_4^+ = bac; & H_6^+ = a; & H_8^+ = a. \end{array}$$

Obviously, H^+ is a solution of Ω^*. Notice, however, that H^+ is not a solution of Ω. Indeed, for the equation $h_1h_2h_3h_4 = h_4h_5h_6h_7$, the word $H_1^+ H_2^+ H_3^+ H_4^+$ is not geodesic as written.

We construct the system of equations $\mathbf{S}$ over $\mathbb{G}$. The system $\mathbf{S}$ consists of all the equations of Ω that correspond to the bases that belong to the periodic structure $\langle \mathcal{P}, R \rangle$, see Example 6.2, and the equation $h(\mathfrak{c}_{e_7}) = h_1 h_2 h_7 = 1$:
$$\mathbf{S} = \{h_1 h_2 h_3 h_4 = h_4 h_5 h_6 h_7,\ h_1 = h_5,\ h_3 = h_7,\ h_1 h_2 h_7 = 1\}.$$

In the example given in Section 6.4, we showed that $H^+(\mathfrak{c}_{e_7}) = 1$ and therefore, H^+ is a solution of $\mathbf{S}$. The cancellation scheme for the solution H^+ is shown on Figure 1.

Notice that the partition table corresponding to the cancellation scheme shown on Figure 1 satisfies condition (7.2).

We then construct the generalised equation $\Upsilon_\mathcal{T}$ associated to the system $\mathbf{S}$ and the generalised equation $\tilde{\Upsilon}_v$ obtained from Υ_v by removing all bases and items that belong to $\mathcal{P}$. Using these two equations we construct $\Omega(\mathcal{P}, R, \mathfrak{c}, \mathcal{T})$ as shown on Figure 2.

We would like to draw the reader's attention to the fact that, though (for simplicity) the generalised equation $\Upsilon_\mathcal{T}$ shown on Figure 2 has not been constructed using the procedure described in Section 3.2.2, but the constructed generalised equation is $\approx$-equivalent to the one described there.

8.3. The extension tree $T_{\text{ext}}(\Omega)$

Recall that the coordinate group $\mathbb{G}_{R(\Omega_v^*)}$ associated to a final leaf v of $T_{\text{dec}}(\Omega)$ is either a proper epimorphic image of $\mathbb{G}_{R(\Omega^*)}$ or all the sections of the corresponding generalised equation Ω_v are non-active constant sections. Since partially commutative groups are equationally Noetherian and thus any sequence of proper epimorphisms of coordinate groups is finite, an inductive argument for those leaves of $T_{\text{dec}}(\Omega)$ that are proper epimorphic images of $\mathbb{G}_{R(\Omega^*)}$ shows that we can construct a tree T_{ext} with the property that for every leaf v of T_{ext} all the sections of the generalised equation Ω_v are non-active constant sections.

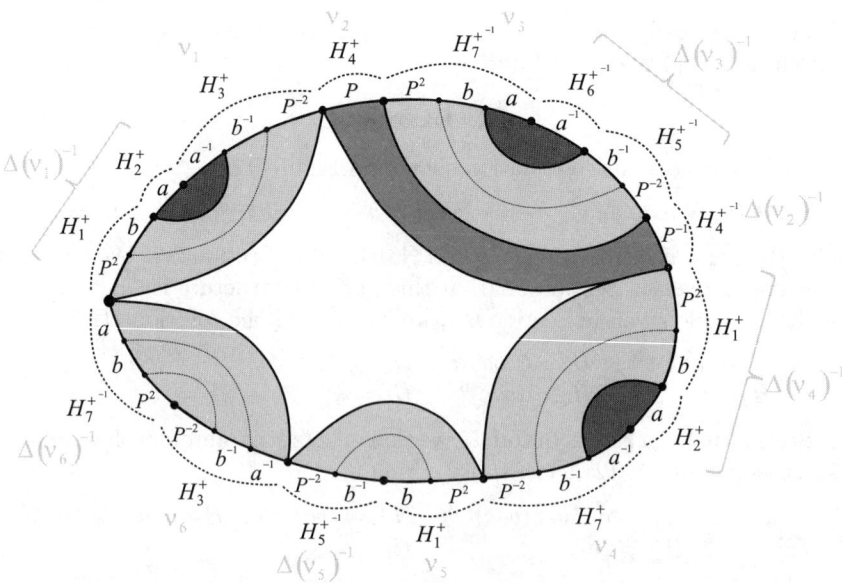

FIGURE 1. The cancellation scheme of H^+.

We summarise the results of this section in the proposition below.

PROPOSITION 8.2. *For a (constrained) generalised equation $\Omega = \Omega_{v_0}$, one can effectively construct a finite oriented rooted at v_0 tree T_{ext}, $T_{\text{ext}} = T_{\text{ext}}(\Omega_{v_0})$, such that:*

(1) *The tree $T_{\text{dec}}(\Omega)$ is a subtree of the tree T_{ext}.*
(2) *To every vertex v of T_{ext} we assign a recursive group of automorphisms $A(\Omega_v)$.*
(3) *For any solution H of a generalised equation Ω there exists a leaf u of the tree T_{ext}, $\text{tp}(u) = 1, 2$, and a solution $H^{[u]}$ of the generalised equation Ω_u such that*
  - $\pi_H = \sigma_0 \pi(v_0, v_1) \sigma_1 \ldots \pi(v_{n-1}, u) \sigma_n \pi_{H^{[u]}}$, *where $\sigma_i \in A(\Omega_{v_i})$;*
  - *the sections of Ω_u are non-active constant sections.*

We define a new transformation L_v, which we call a *leaf-extension of the tree $T_{\text{dec}}(\Omega)$ at the leaf v* in the following way. If there are active sections in the generalised equation Ω_v, we take the union of two trees $T_{\text{dec}}(\Omega)$ and $T_{\text{dec}}(\Omega_v)$ and identify the leaf v of $T_{\text{dec}}(\Omega)$ with the root v of the tree $T_{\text{dec}}(\Omega_v)$, i.e. we extend the tree $T_{\text{dec}}(\Omega)$ by gluing the tree $T_{\text{dec}}(\Omega_v)$ to the vertex v. If all the sections of the generalised equation Ω_v are non-active, then the vertex v is a leaf and $T_{\text{dec}}(\Omega_v)$ consists of a single vertex, namely v. We call such a vertex v *terminal*.

We use induction to construct the extension tree $T_{\text{ext}}(\Omega)$. Let v be a non-terminal leaf of $T^{(0)} = T_{\text{dec}}(\Omega)$. Apply the transformation L_v to obtain a new tree $T^{(1)} = L_v(T_{\text{dec}}(\Omega))$. If in $T^{(1)}$ there exists a non-terminal leaf v_1, we apply the transformation L_{v_1}, and so on. By induction we construct a strictly increasing sequence of finite trees

(8.2) $$T^{(0)} \subset T^{(1)} \subset \ldots \subset T^{(i)} \subset \ldots$$

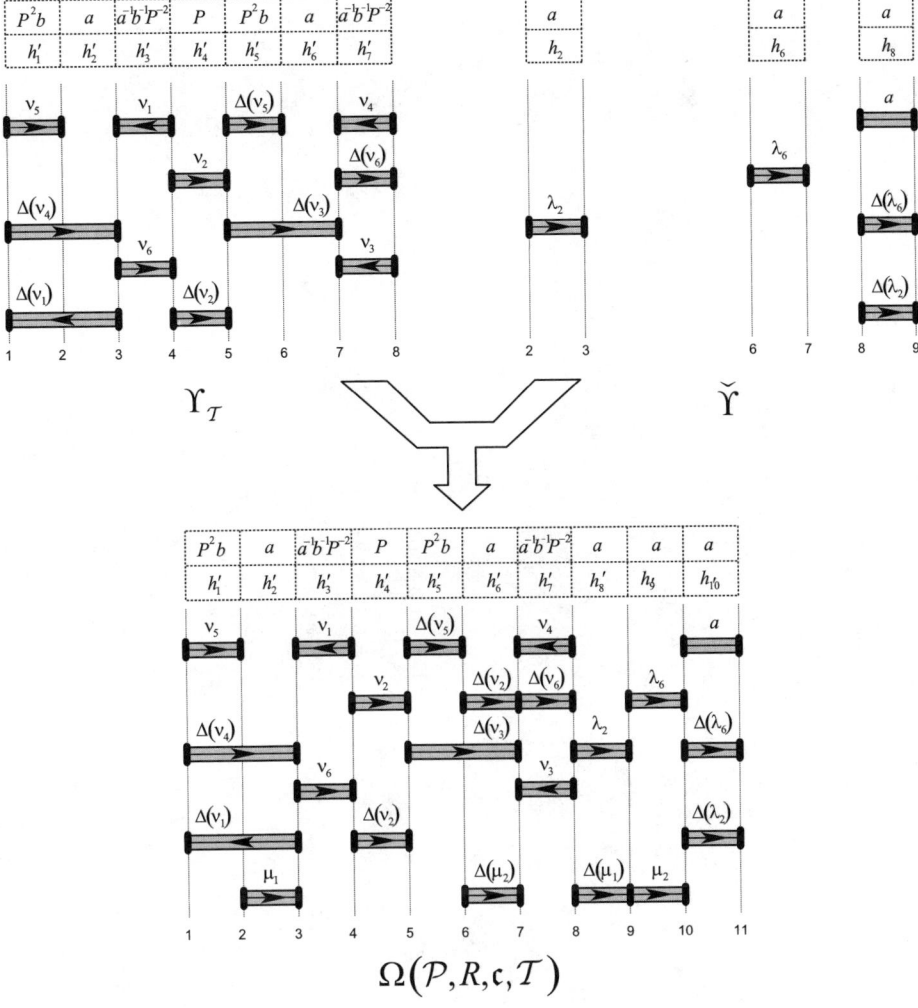

FIGURE 2. The generalised equations $\Upsilon_{\mathcal{T}}$, $\check{\Upsilon}$ and $\Omega(\mathcal{P}, R, \mathfrak{c}, \mathcal{T})$.

Sequence (8.2) is finite. Indeed, assume the contrary, i.e. the sequence is infinite and hence the union $T^{(\infty)}$ of this sequence is an infinite, locally finite tree. By König's lemma, $T^{(\infty)}$ has an infinite branch. Observe that along any infinite branch in $T^{(\infty)}$ one has to encounter infinitely many proper epimorphisms. This derives a contradiction with the fact that $\mathbb{G}$ is equationally Noetherian.

Denote by $T_{\text{ext}}(\Omega)$ the last term of the sequence (8.2).

The groups of automorphisms $A(\Omega_v)$ associated to vertices of $T_{\text{ext}}(\Omega)$ are induced, in a natural way, by the groups of automorphisms associated to vertices of the decomposition trees.

Naturally, we call leaves u such that the paths $\mathfrak{p}(H)$ associated to solutions H of Ω end in u, *final leaves of $T_{\text{ext}}(\Omega)$*.

CHAPTER 9

The solution tree $T_{\text{sol}}(\Omega)$ and the main theorem

In the previous chapter we have shown that the generalised equations $\Omega_v = \langle \Upsilon_v, \Re_{\Upsilon_v} \rangle$ associated to the final leaves v of $T_{\text{ext}}(\Omega)$ contain only non-active constant sections. In other words, the coordinate group $\mathbb{G}_{R(\Omega_v^*)}$ is isomorphic to

$$\mathbb{G}[h_1, \ldots, h_{\rho\Omega_v}]/R(\{\text{coefficient equations}\} \cup \{[h_i, h_j] \mid \Re_{\Upsilon_v}(h_i, h_j)\}).$$

In general, the coordinate group $\mathbb{G}_{R(\Omega_v^*)}$ may be not fully residually $\mathbb{G}$, see the example given in Section 9.1.

The solution tree $T_{\text{sol}}(\Omega)$ is constructed from the tree $T_{\text{ext}}(\Omega)$ by adding some edges, labelled by homomorphisms of the coordinate groups, to final leaves of the tree T_{ext}. The new vertices $w_1, \ldots, w_n$ connected to a final leaf v of T_{ext} by an edge have the coordinate groups $\mathbb{G}_{w_i}$ associated to them and satisfy the following properties.

(1) For every solution $H^{(v)}$ of Ω_v, there exist a vertex w_i and a solution $H^{(w_i)}$ such that the following diagram commutes:

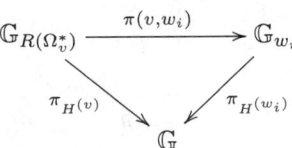

(2) For every solution solution $H^{(w_i)}$ there exists a homomorphism $\varphi : \mathbb{G}_{R(\Omega_v^*)} \to \mathbb{G}$ such that the following diagram commutes:

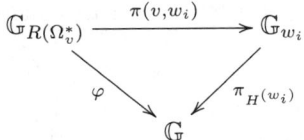

(3) The coordinate group $\mathbb{G}_{w_i}$ is a fully-residually $\mathbb{G}$ free partially commutative group.

The idea behind this construction is the following.

Let v be a final leaf of the tree T_{ext} and Ω_v be the constrained generalised equation associated to v. By definition, for any $h_i, h_j \in h^{(v)} = h$, such that $\Re_{\Upsilon_v}(h_i, h_j)$, one has that $H_i^{(v)} \leftrightarrows H_j^{(v)}$ for any solution $H^{(v)}$ of Ω_v, i.e. $H_i^{(v)} \in \mathbb{A}(H_j^{(v)})$ and $H_j^{(v)} \in \mathbb{A}(H_i^{(v)})$, see Section 2.5. Therefore, any solution of the generalised equation maps h_i and h_j into disjoint canonical parabolic subgroups of $\mathbb{G}$ that $\leftrightarrows$-commute.

We encode all canonical parabolic subgroups of the group $\mathbb{G}$ and the $\leftrightarrows$-commutativity relation between them in the graph F. There are only finitely many tuples of canonical parabolic subgroups, where the solution $H^{(v)}$ of Ω_v may map the tuple of variables h. The choice of the tuple of canonical parabolic subgroups is encoded by the graph homomorphism $\varphi_{v,i}$ defined below. There is a vertex w_i in the tree T_{sol} for every such homomorphism $\varphi_{v,i}$.

Using the homomorphism $\varphi_{v,i}$, we construct the coordinate group $\mathbb{G}_{w_i}$ that we associate to the vertex w_i and define solutions corresponding to this vertex.

We then prove that properties (1), (2) and (3) above hold.

We refer the reader to Section 9.1 for an example of the constructions described in this chapter.

Define a non-oriented graph F as follows. The set of vertices $V(F)$ is the set of all full subgraphs of the commutation graph $\mathcal{G}$ of $\mathbb{G}$ (or, which is equivalent, canonical parabolic subgroups of the group $\mathbb{G}$). There is an edge $e \in E(F)$ between two vertices s_1 and s_2 if and only $V(s_1) \leftrightarrows V(s_2)$, $V(s_1), V(s_2) \subseteq \mathcal{A}$.

Let Ω_v be the generalised equation associated to a final leaf v of the tree T_{ext}. Define a non-oriented graph Π_v as follows. The set of vertices $V(\Pi_v)$ is $\mathcal{A} \cup \{h_i \in h^{(v)} \mid h_i \text{ does not occur in coefficient equations of } \Omega_v\}$. There is an edge $e \in E(\Pi_v)$ between two vertices v_1 and v_2 if and only

$v_1 = a_i$, $v_2 = a_j$ and $[a_i, a_j] = 1$ in $\mathbb{G}$,
$v_1 = h_i$, $v_2 = h_j$ and $\Re_{\Upsilon_v}(h_i, h_j)$,
$v_1 = h_i$, $v_2 = a_j$ and there exists an equation $h_k^{\pm 1} = a_j$, and $\Re_{\Upsilon_v}(h_i, h_k)$.

Consider the set of all graph homomorphisms $\varphi_{v,i}$ from Π_v to F that satisfy the following conditions:

(I) $\varphi_{v,i}(a_j) = \{a_j\}$, for all $a_j \in \mathcal{A}$;
(II) for every $e \in E(\Pi_v)$ we have $\varphi_{v,i}(e) \in E(F)$, i.e. $\varphi_{v,i}(e)$ does not collapse edges.

Note that the set of all such homomorphisms is finite and can be effectively constructed.

For every homomorphism $\varphi_{v,i}$ we construct a new leaf w_i in the tree $T_{\text{sol}}(\Omega)$ and an edge joining v and w_i.

The coordinate group $\mathbb{G}_{w_i}$ associated to the vertex w_i is obtained as follows. Let G be the graph product of groups with the underlying commutation graph $\varphi_{v,i}(\Pi_v)$.

The group associated to the vertex s_j of $\varphi_{v,i}(\Pi_v)$ is defined as follows. For every vertex s_j of $\varphi_{v,i}(\Pi_v)$ we consider the partially commutative group $\mathbb{G}(s_j)$, where $\mathbb{G}(s_j) < \mathbb{G}$ is the canonical parabolic subgroup of $\mathbb{G}$ corresponding to the full subgraph of $\mathcal{G}$ associated to s_j. Consider the decomposition of $\mathbb{G}(s_j)$ of the form (2.1):

$$\mathbb{G}(s_j) = \mathbb{G}(s_j, I_1) \times \cdots \times \mathbb{G}(s_j, I_{m_j}).$$

Let $h^{(j)}$ be the set of vertices $h_k \in V(\Pi_v)$ such that $\varphi_{v,i}(h_k) = s_j$. We treat every $h_k \in h^{(j)}$ as a tuple of variables $(h_{k,1}, \ldots, h_{k,m_j})$, $h_{k,l} \in h^{(j,I_l)}$. Consider the group

(9.1) $$\mathbb{G}(s_j, I_1)[h^{(j,I_1)}] \times \cdots \times \mathbb{G}(s_j, I_m)[h^{(j,I_{m_j})}],$$

where if $\mathbb{G}(s_j, I_l)$ is free abelian, then $\mathbb{G}(s_j, I_l)[h^{(j,I_l)}] = \mathbb{G}(s_j, I_l) \times \langle h^{(j,I_l)} \rangle$, where $\langle h^{(j,I_l)} \rangle$ is the free abelian group with basis $h^{(j,I_l)}$, and if $\mathbb{G}(s_j, I_l)$ is non-abelian, then $\mathbb{G}(s_j, I_l)[h^{(j,I_l)}] = \mathbb{G}(s_j, I_l) * F(h^{(j,I_l)})$. The group we associate to the vertex

9. THE SOLUTION TREE $T_{sol}(\Omega)$ AND THE MAIN THEOREM

s_j is the group given in (9.1). It follows, since the graph product of partially commutative groups is again a partially commutative group, that G is a free partially commutative group.

We now turn the group G into a $\mathbb{G}$-group as follows

$$\mathbb{G}_{w_i} = \left\langle \mathbb{G}, G \,\middle|\, C, [C_\mathbb{G}(\mathbb{G}(s_j, I_k)), h^{(j,I_k)}] = 1 \text{ for all } j, k \right\rangle,$$

where the relations in C identify the subgroups $\mathbb{G}(s_j)$ with the corresponding subgroups of $\mathbb{G}$. This is the group $\mathbb{G}_{w_i}$ that we associate to the leaf w_i of the tree $T_{ext}(\Omega)$. Note that, since G is a partially commutative group and the centraliser of a canonical parabolic subgroup is again a canonical parabolic subgroup, the group $\mathbb{G}_{w_i}$ is a free partially commutative group.

A $\mathbb{G}$-homomorphism ψ from $\mathbb{G}_{w_i}$ to $\mathbb{G}$ such that for every $h_k \in h^{(j)}$ one has $\psi(h_k) \in \mathbb{G}(s_j)$ is termed a *solution associated to the vertex* w_i. In other words, by taking restrictions, every solution associated to the vertex w_i, induces a tuple of homomorphisms $(\psi_1, \ldots, \psi_{|V(\varphi_{v,i}(\Pi_v))|})$, where ψ_j is a $\mathbb{G}(s_j)$-homomorphism from $\mathbb{G}(s_j)[h^{(j)}]$ to $\mathbb{G}(s_j)$. Conversely, any $|V(\varphi_{v,i}(\Pi_v))|$-tuple of $\mathbb{G}(s_j)$-homomorphisms from $\mathbb{G}(s_j)[h^{(j)}]$ to $\mathbb{G}(s_j)$ uniquely defines a solution ψ associated to the vertex w_i. The homomorphisms ψ_j are called *components* of ψ.

Every homomorphism ψ_j, in turn, induces an m_j-tuple of $\mathbb{G}(s_j, I_l)$-homomorphisms from $\mathbb{G}(s_j, I_l)[h^{(j,I_l)}]$ to $\mathbb{G}(s_j, I_l)$, $l = 1, \ldots, m_j$. Conversely, any m_j-tuple of $\mathbb{G}(s_j, I_l)$-homomorphisms from $\mathbb{G}(s_j, I_l)[h^{(j,I_l)}]$ to $\mathbb{G}(s_j, I_l)$ uniquely defines a component ψ_j of a solution ψ.

The map $h_k \mapsto h_{k,1} \cdots h_{k,m_j}$ induces a $\mathbb{G}$-homomorphism $\pi(v, w_i)$ from $\mathbb{G}_{R(\Omega_v^*)}$ to $\mathbb{G}_{w_i}$. Indeed, by Proposition 9.1 below, $\mathbb{G}_{w_i}$ is $\mathbb{G}$-discriminated by $\mathbb{G}$. Then, by Theorem 2.4, $\mathbb{G}_{w_i}$ is the coordinate group of an irreducible algebraic set. Since every equation from the system Ω_v^* is mapped to the trivial element of $\mathbb{G}_{w_i}$ and since $\mathbb{G}_{w_i}$ is a coordinate group, $\pi(v, w_i)$ is a homomorphism.

It follows, therefore, that property (2) described in the beginning of this section holds, i.e. that for every solution solution $H^{(w_i)}$ there exists a homomorphism $\varphi: \mathbb{G}_{R(\Omega_v^*)} \to \mathbb{G}$ such that the following diagram commutes:

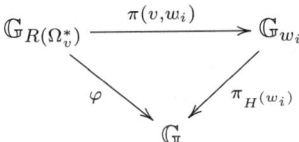

It suffices to take $\varphi: \mathbb{G}_{R(\Omega_v^*)} \to \mathbb{G}$ to be the composition of $\pi(v, w_i)$ and $\pi_{H^{(w_i)}}$.

We now prove that property (1) described in the beginning of this chapter holds, i.e. that for every solution H of Ω_v there exists i and a solution ψ associated to the vertex w_i such that the following diagram is commutative:

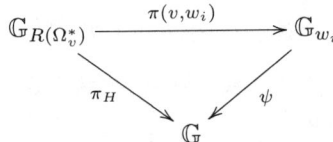

Given a solution H of Ω_v, we construct a non-oriented graph $\mathbb{J}$. There are two types of vertices in $\mathbb{J}$. For every $a_j \in \mathcal{A}$, we add a vertex labelled by a_j. For all distinct sets $\text{alph}(H_j)$, we introduce a vertex of $\mathbb{J}$ labelled by a full subgraph of $\mathcal{G}$ generated by alph(H_j). There is an edge between two vertices corresponding to the sets alph(H_j) and alph(H_k) if and only if alph(H_j) ⇆ alph(H_k). There is an edge between two vertices corresponding to a_j and a_k if and only if a_j ⇆ a_k. There is an edge between two vertices v_1, corresponding to the set alph(H_j), and v_2, corresponding to a_k, if and only if alph(H_j) ⇆ a_k. By construction, the graph ℑ is a full subgraph of F.

The map $\phi_{v,i} : h_j \mapsto v(h_j)$, where $v(h_j)$ is labelled by the full subgraph of $\mathcal{G}$ generated by alph(H_j) and $\phi_{v,i} : a_j \mapsto a_j$, where $a_j \in \mathcal{A}$ extends, to an epimorphism φ from Π_v to ℑ. Since H is a solution of Ω_v, if $\Re_{\Upsilon_v}(h_j, h_k)$, then H_j ⇆ H_k. Thus, the homomorphism φ satisfies the property (II) above and hence, $\varphi = \varphi_{v,i}$ for some i. Setting $\psi = \pi_H$, it follows that the above diagram is commutative.

The proposition below proves that property (3) described in the beginning of this chapter holds.

PROPOSITION 9.1. *In the above notation, the groups* $\mathbb{G}_{w_i}$ *are* $\mathbb{G}$-*discriminated by* $\mathbb{G}$ *by the family of solutions associated to the vertex* w_i.

PROOF. We use induction on the number N of vertices of the underlying graph $\varphi_{v,i}(\Pi_v)$ of the graph product G to prove that the group $\mathbb{G}_{w_i}$ is obtained from $\mathbb{G}$ by a chain of extension of centralisers of directly indecomposable canonical parabolic subgroups. It then follows, by Proposition 2.12, that the group $\mathbb{G}_{w_i}$ is $\mathbb{G}$-discriminated by $\mathbb{G}$.

If $N = 1$, then

$$\mathbb{G}_{w_1} = \left\langle \begin{array}{l} \mathbb{G}, \mathbb{G}(s_1, I_1)[h^{(1,I_1)}] \times \cdots \times \mathbb{G}(s_1, I_{m_1})[h^{(1,I_{m_1})}] \\ C, [C_\mathbb{G}(\mathbb{G}(s_1, I_k)), h^{(1,I_k)}] = 1 \text{ for all } k \end{array} \right\rangle,$$

where the relations in C identify the subgroups $\mathbb{G}(s_1, I_k)$ with the corresponding subgroups of $\mathbb{G}$.

We use induction on m_1 to prove that the group $\mathbb{G}_{w_1}$ is obtained from $\mathbb{G}$ by a chain of extension of centralisers of directly indecomposable canonical parabolic subgroups. If $m_1 = 1$, then the statement is trivial.

Let

$$\mathbb{G}'_{w_1} = \left\langle \begin{array}{l} \mathbb{G}, \mathbb{G}(s_1, I_1)[h^{(1,I_1)}] \times \cdots \times \mathbb{G}(s_1, I_{m_1-1})[h^{(1,I_{m_1-1})}], \\ C, [C_\mathbb{G}(\mathbb{G}(s_1, I_k)), h^{(1,I_k)}] = 1 \text{ for all } k \end{array} \right\rangle.$$

Without loss of generality we may assume that $|h^{(1,I_{m_1})}| = 1$, $h^{(1,I_{m_1})} = \{h\}$. It suffices to show that $\mathbb{G}_{w_1}$ is isomorphic to

(9.2) $\qquad \langle \mathbb{G}'_{w_1}, h \mid \text{rel}(\mathbb{G}'_{w_1}), [h, C_{\mathbb{G}'_{w_1}}(\mathbb{G}(s_1, I_{m_1}))] = 1 \rangle.$

Since, by induction, $\mathbb{G}'_{w_1}$ is a partially commutative group, by Theorem 2.5,

$$C_{\mathbb{G}'_{w_1}}(\mathbb{G}(s_1, I_{m_1})) = \langle C_\mathbb{G}(\mathbb{G}(s_1, I_{m_1})), h^{(1,I_k)}, k = 1, \ldots, m_1 - 1 \rangle.$$

Comparing the presentations of $\mathbb{G}_{w_1}$ and the one given in (9.2), the statement follows in the case when $N = 1$.

Take a graph Λ with N vertices and consider a full subgraph Λ' of Λ such that $V(\Lambda) = V(\Lambda') \cup \{s_N\}$. By induction, the $\mathbb{G}$-group $\mathbb{G}_{\Lambda'}$ constructed by the graph Λ' is a partially commutative group obtained from $\mathbb{G}$ by a sequence of extensions of centralisers of directly indecomposable canonical parabolic subgroups.

Suppose first that $\mathbb{G}(s_N)$ is directly indecomposable. Without loss of generality we may assume that $|h^{(N,I_1)}| = 1$, $h^{(N,I_1)} = \{h_N\}$. We now prove that $\mathbb{G}_\Lambda$ is isomorphic to
$$\mathbb{G}' = \langle \mathbb{G}_{\Lambda'}, h_N \mid \mathrm{rel}(\mathbb{G}_{\Lambda'}), [C_{\mathbb{G}_{\Lambda'}}(\mathbb{G}(s_N, I_1)), h_N] = 1 \rangle.$$
Since, by induction, $\mathbb{G}_{\Lambda'}$ is a partially commutative group, by Theorem 2.5,
$$C_{\mathbb{G}_{\Lambda'}}(\mathbb{G}(s_N, I_1)) = \left\langle \begin{array}{c} C_\mathbb{G}(\mathbb{G}(s_N, I_1)),\ h^{(j,I_k)},\ k = 1,\ldots,m_j \\ \text{and there is an edge between } s_j \text{ and } s_N \text{ in } \Lambda \end{array} \right\rangle$$
Comparing the presentations of $\mathbb{G}'$ and $\mathbb{G}_\Lambda$, the statement follows.

If the group $\mathbb{G}(s_N)$ is directly decomposable, the proof is analogous to the base of induction and the previous case. □

The solution tree $T_{\mathrm{sol}}(\Omega)$ is obtained from the extension tree $T_{\mathrm{ext}}(\Omega)$ by extending every final leaf of $T_{\mathrm{ext}}(\Omega)$ as above. To each non-leaf vertex v of the tree $T_{\mathrm{sol}}(\Omega)$, we associate the same group of automorphisms $A(\Omega_v)$ as in the extension tree $T_{\mathrm{ext}}(\Omega)$. For every leaf w_i of $T_{\mathrm{sol}}(\Omega)$ we associate the trivial group of automorphisms.

We are now ready to formulate the main result of this paper.

THEOREM 9.2. *Let $\Omega = \Omega(h)$ be a constrained generalised equation in variables h. Let $T_{\mathrm{sol}}(\Omega)$ be the solution tree for Ω. Then the following statements hold.*

(1) *For any solution H of the generalised equation Ω there exist: a path $v_0 \to v_1 \to \ldots \to v_n = v$ in $T_{\mathrm{sol}}(\Omega)$ from the root vertex v_0 to a leaf v, a sequence of automorphisms $\sigma = (\sigma_0, \ldots, \sigma_n)$, where $\sigma_i \in A(\Omega_{v_i})$ and a solution $H^{(v)}$ associated to the vertex v, such that*

(9.3) $$\pi_H = \Phi_{\sigma, H^{(v)}} = \sigma_0 \pi(v_0, v_1) \sigma_1 \ldots \pi(v_{n-1}, v_n) \sigma_n \pi_{H^{(v)}}.$$

(2) *For any path $v_0 \to v_1 \to \ldots \to v_n = v$ in $T_{\mathrm{sol}}(\Omega)$ from the root vertex v_0 to a leaf v, any sequence of automorphisms $\sigma = (\sigma_0, \ldots, \sigma_n)$, $\sigma_i \in A(\Omega_{v_i})$, and any solution $H^{(v)}$ associated to the vertex v, the homomorphism $\Phi_{\sigma, H^{(v)}}$ is a solution of Ω^*. Moreover, every solution of Ω^* can be obtained in this way.*

PROOF. The statement follows from the construction of the tree $T_{\mathrm{sol}}(\Omega, \Lambda)$. □

We call leaves v of the tree $T_{\mathrm{sol}}(\Omega)$ such that there exists a path from v_0 to v as in statement (1) of Theorem 9.2 *final leaves of the tree* $T_{\mathrm{sol}}(\Omega)$.

THEOREM 9.3. *Let $\mathbb{G}$ be the free partially commutative group with the underlying commutation graph $\mathcal{G}$ and let $\widehat{G}$ be a finitely generated ($\mathbb{G}$-)group. Then the set of all ($\mathbb{G}$-)homomorphisms $\mathrm{Hom}(\widehat{G}, \mathbb{G})$ ($\mathrm{Hom}_\mathbb{G}(\widehat{G}, \mathbb{G})$, correspondingly) from $\widehat{G}$ to $\mathbb{G}$ can be effectively described by a finite rooted tree. This tree is oriented from the root, all its vertices except for the root and the leaves are labelled by coordinate groups of generalised equations. The final leaves of the tree are labelled by fully residually $\mathbb{G}$ partially commutative groups $\mathbb{G}_{w_i}$.*

Edges from the root vertex correspond to a finite number of ($\mathbb{G}$-)homomorphisms from $\widehat{G}$ into coordinate groups of generalised equations. To each vertex group we assign the group of automorphisms $A(\Omega_v)$. Each edge (except for the edges from the root and the edges to the final leaves) in this tree is labelled by an epimorphism, and all the epimorphisms are proper. Every ($\mathbb{G}$-)homomorphism from $\widehat{G}$ to $\mathbb{G}$ can be written as a composition of the ($\mathbb{G}$-)homomorphisms corresponding to the edges, automorphisms of the groups assigned to the vertices, and a ($\mathbb{G}$-)homomorphism $\psi = (\psi_j)_{j \in J}$, $|J| \leq 2^{\mathfrak{r}}$ into $\mathbb{G}$, where $\psi_j : \mathbb{H}_j[Y] \to \mathbb{H}_j$ and $\mathbb{H}_j$ is the free partially commutative subgroup of $\mathbb{G}$ defined by some full subgraph of $\mathcal{G}$.

PROOF. Suppose first that $\widehat{G}$ is the finitely generated $\mathbb{G}$-group $\mathbb{G}$-generated by X, i.e. $\widehat{G}$ is generated by $\mathbb{G} \cup X$. Let S be the set of defining relations of $\widehat{G}$. We treat S as a system of equations (possibly infinite) over $\mathbb{G}$ (with coefficients from $\mathbb{G}$). Since $\mathbb{G}$ is equationally Noetherian, there exists a finite subsystem $S_0 \subseteq S$ such that $R(S) = R(S_0)$. Since every $\mathbb{G}$-homomorphism from $\widehat{G} = \langle \mathbb{G}, X \rangle / \mathrm{ncl}\langle S \rangle$ to $\mathbb{G}$ factors through $\mathbb{G}_{R(S_0)}$, it suffices to describe the set of all homomorphisms from $\mathbb{G}_{R(S_0)}$ to $\mathbb{G}$. We now run the process for the system of equations S_0. Since there is a one-to-one correspondence between solutions of the system S_0 and homomorphisms from $\mathbb{G}_{R(S_0)}$ to $\mathbb{G}$, by Theorem 9.2, we obtain a description of $\mathrm{Hom}_{\mathbb{G}}(\mathbb{G}_{R(S_0)}, \mathbb{G})$.

Suppose now that $\widehat{G}$ is not a $\mathbb{G}$-group, $\widehat{G} = \langle X \mid S \rangle$ (note that the set S may be infinite). We treat S as a coefficient-free system of equations over $\mathbb{G}$. Though, formally, coordinate groups are $\mathbb{G}$-groups, in this case, we consider instead the group $\mathbb{G}'_{R'(S)} = F(X) / R'(S)$, where $R'(S) = \bigcap \ker(\varphi)$ and the intersection is taken over all homomorphisms φ from $F(X)$ to $\mathbb{G}$ such that $S \subseteq \ker(\varphi)$. It is clear that $\mathbb{G}'_{R'(S)}$ is a residually $\mathbb{G}$ group. Since $\mathbb{G}$ is equationally Noetherian, there exists a finite subsystem $S_0 \subseteq S$ such that $R'(S) = R'(S_0)$. Every homomorphism from $\widehat{G}$ to $\mathbb{G}$ factors through $\mathbb{G}'_{R'(S_0)}$, it suffices to describe the homomorphisms from $\mathbb{G}'_{R'(S_0)}$ to $\mathbb{G}$. We now run the process for the system of equations S_0 and construct the solution tree, where to each vertex v instead of $\mathbb{G}_{R(\Omega_v^*)}$ we associate $\mathbb{G}'_{R'(\Omega_v^*)}$ and the corresponding epimorphisms, homomorphisms and automorphisms are defined in a natural way. We thereby obtain a description of the set $\mathrm{Hom}(\mathbb{G}'_{R'(S_0)}, \mathbb{G})$. □

9.1. Example

Let $\mathbb{G}$ be the partially commutative group whose commutation graph is a path of length 3, $\mathbb{G} = \langle a, b, c, d \mid [a, b] = 1, [b, c] = 1, [c, d] = 1 \rangle$. Consider the following coordinate group, that, a priori, could be associated to the leaf of the extension tree:

$$\mathbb{G}_{R(\Omega_v^*)} = \mathbb{G}[h_1, \ldots, h_6] \Big/ R \left(\begin{array}{l} \{h_5 = a,\ h_6 = c\} \cup \\ \{[h_1, h_2] = [h_2, h_3] = [h_3, h_4] = 1 \\ [h_1, h_4] = [h_1, h_5] = [h_4, h_6] = 1 \end{array} \right\} \right).$$

We show that the group $\mathbb{G}_{R(\Omega_v^*)}$ is not fully residually $\mathbb{G}$. We first show that the elements $[h_1, h_3]$ and $[h_2, h_4]$ are non-trivial in $\mathbb{G}_{R(\Omega_v^*)}$. By the definition of the radical, it suffices to show that there exist homomorphisms ψ_1, ψ_2 from $\mathbb{G}_{R(\Omega_v^*)}$ to $\mathbb{G}$ such that $\psi_1([h_1, h_3]) \neq 1$ and $\psi_2([h_2, h_4]) \neq 1$. It is easy to check that the map $h_1 \to a$, $h_2 \to b$, $h_3 \to c$ and $h_4 \to b$ induces a homomorphism ψ_1 such that $\psi_1([h_1, h_3]) \neq 1$. Similarly, the map $h_1 \to b$, $h_2 \to a$, $h_3 \to b$ and $h_4 \to c$ induces a homomorphism ψ_2 such that $\psi_2([h_2, h_4]) \neq 1$.

We claim that the family $\{[h_1, h_3], [h_2, h_4]\}$ of elements from $\mathbb{G}_{R(\Omega_v^*)}$ can not be discriminated into $\mathbb{G}$, i.e. for every homomorphism ψ from $\mathbb{G}_{R(\Omega_v^*)}$ to $\mathbb{G}$ either $\psi([h_1, h_3]) = 1$ or $\psi([h_2, h_4]) = 1$. In the proof we make a substantial use of the Centraliser Theorem, see Theorem 2.5.

9.1. EXAMPLE

Indeed, since $[h_1, h_5] = 1$ and $h_5 = a$, we have that $\psi(h_1) \in C_\mathbb{G}(a)$. There are three cases to consider: either $\psi(h_1) \in \langle a \rangle$ or $\psi(h_1) \in \langle a, b \rangle$ (and $\psi(h_1) \notin \langle a \rangle \cup \langle b \rangle$) or $\psi(h_1) \in \langle b \rangle$.

Suppose that $\psi(h_1) \in \langle a \rangle$ (or, $\psi(h_1) \in \langle a, b \rangle$, $\psi(h_1) = w(a, b)$). Since $\psi(h_4) \in C_\mathbb{G}(c) \cap C_\mathbb{G}(a)$ (correspondingly, $\psi(h_4) \in C_\mathbb{G}(c) \cap C_\mathbb{G}(w)$), it follows that $\psi(h_4) \in \langle b \rangle$. Similarly, we get that $\psi(h_2) \in C_\mathbb{G}(a) = \langle a, b \rangle$, and therefore $\psi([h_2, h_4]) = 1$.

Suppose that $\psi(h_1) \in \langle b \rangle$. Then either $\psi(h_4) \in \langle b \rangle$ or $\psi(h_4) \in \langle c \rangle$ or $\psi(h_4) \in \langle b, c \rangle$. If $\psi(h_4) \in \langle b \rangle$ or $\psi(h_4) \in \langle b, c \rangle$, then $h_3 \in C_\mathbb{G}(b)$ and thus $\psi([h_1, h_3]) = 1$. Finally, assume that $\psi(h_4) \in \langle c \rangle$. It follows that either $\psi(h_3) \in \langle c \rangle$ or $\psi(h_3) \in \langle d \rangle$ or $\psi(h_4) \in \langle c, d \rangle$. If $\psi(h_3) \in \langle c \rangle$, then $\psi([h_1, h_3]) = 1$. If $\psi(h_3) \in \langle c \rangle$ or $\psi(h_3) \in \langle c, d \rangle$, then since $\psi(h_2) \in C_\mathbb{G}(\psi(h_1)) \cap C_\mathbb{G}(\psi(h_3))$, it follows that $\psi(h_2) \in \langle c \rangle$, therefore, $\psi([h_2, h_4]) = 1$.

Therefore, the group $\mathbb{G}_{R(\Omega_v^*)}$ is not $\mathbb{G}$-discriminated by $\mathbb{G}$.

Note that the set of solutions of the generalised equation Ω_v is empty, while there are solutions of the system Ω_v^*. Indeed, if H is a solution of Ω_v, then $H_1 \leftrightarrows a$, therefore $H_1 \in \langle b \rangle$. On the other hand, since $H_4 \leftrightarrows c$, we get that $H_4 \in \langle b, d \rangle$. This derives a contradiction since $H_1 \not\leftrightarrows H_4$.

Let $\mathbb{G}$ be the partially commutative group whose commutation graph is a path of length 3, $\mathbb{G} = \langle a, b, c, d \mid [a, b] = 1, [b, c] = 1, [c, d] = 1 \rangle$. Consider the following coordinate group:

$$\mathbb{G}_{R(\Omega_v^*)} = \mathbb{G}[h_1, \ldots, h_6] \Big/ R \left(\begin{array}{l} \{h_5 = a,\ h_6 = b\} \cup \\ \{[h_1, h_2] = [h_2, h_3] = [h_3, h_4] = 1 \\ \phantom{\{}[h_1, h_4] = [h_1, h_5] = [h_4, h_6] = 1 \end{array} \right\} \right).$$

We now construct one of the groups $\mathbb{G}_{w_i}$. The corresponding graphs Π_v and F constructed by Ω_v and $\mathbb{G}$ are shown on Figure 1.

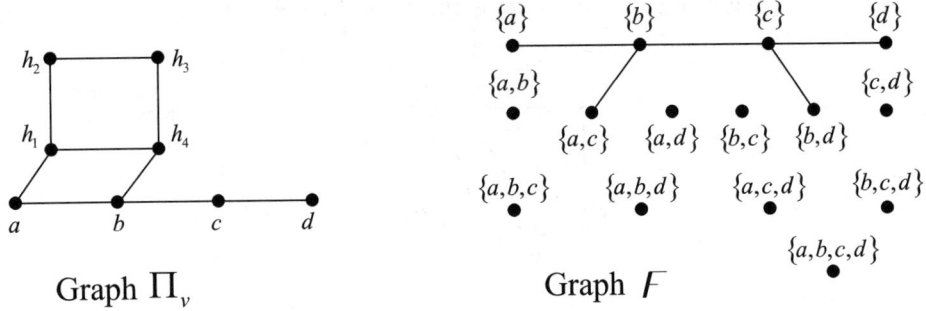

FIGURE 1. The graphs Π_v and F.

The set of all homomorphisms $\varphi_{v,i}$ from Π_v to F that satisfy properties (I) and (II) are listed in the table below:

	$\varphi_{v,1}$	$\varphi_{v,2}$	$\varphi_{v,3}$	$\varphi_{v,4}$	$\varphi_{v,5}$	$\varphi_{v,6}$	$\varphi_{v,7}$	$\varphi_{v,8}$	$\varphi_{v,9}$	$\varphi_{v,10}$	$\varphi_{v,11}$
h_1	b	b	b	b	b	b	b	b	b	b	b
h_2	a	a	a	$\{a,c\}$	$\{a,c\}$	$\{a,c\}$	c	c	c	c	c
h_3	b	b	b	b	b	b	b	b	b	$\{b,d\}$	d
h_4	a	$\{a,c\}$	c	a	$\{a,c\}$	c	a	$\{a,c\}$	c	c	c

Consider the homomorphism $\varphi_{v,6}$ then the corresponding graph $\varphi_{v,6}(\Pi_v)$ is a path of length 2, whose vertices s_1, s_2 and s_3 are labelled by the subgroups of

$\mathbb{G}$ generated by $\{a,c\}$, $\{b\}$ and $\{c\}$ correspondingly, see Figure 2. The partially commutative groups $\mathbb{G}(s_1) = \mathbb{G}(a,c)$, $\mathbb{G}(s_2) = \mathbb{G}(b)$ and $\mathbb{G}(s_3) = \mathbb{G}(c)$ are directly indecomposable (the first one is free, and the latter two are cyclic), thus their decompositions of the form (2.1) are trivial. The set of vertices $h_k \in V(\Pi_v)$ such that $\varphi_{v,6}(h_k) = s_1$ is $h^{(1)} = \{h_2\}$. Analogously, $h^{(2)} = \{h_1, h_3\}$ and $h^{(2)} = \{h_4\}$. The corresponding group G is a graph product whose underlying graph is a path of length two and the corresponding vertex groups are $\mathbb{G}(a,c)[h_2]$, $\langle b \rangle \times \langle h_1 \rangle \times \langle h_3 \rangle$ and $\langle c \rangle \times \langle h_4 \rangle$, see Figure 2.

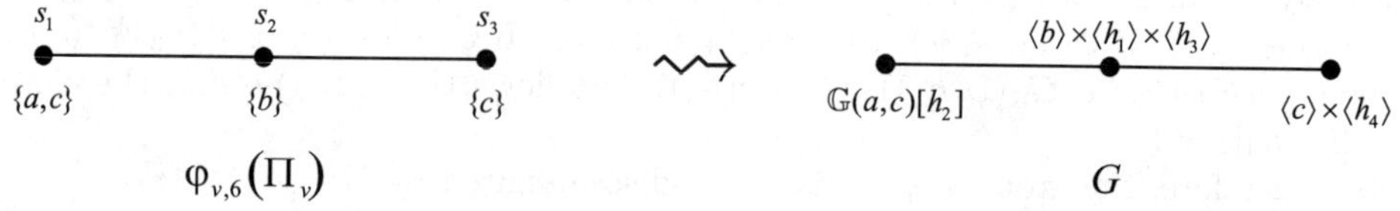

FIGURE 2. Constructing the group G by the graph $\varphi_{v,6}(\Pi_v)$.

We then construct the coordinate group
$$\mathbb{G}_{w_6} = \langle \mathbb{G}, G \mid C, [C_\mathbb{G}(\mathbb{G}(a,c)), h^{(1)}] = 1, [C_\mathbb{G}(\mathbb{G}(b)), h^{(2)}] = 1, [C_\mathbb{G}(\mathbb{G}(c)), h^{(3)}] = 1 \rangle$$
$$= \left\langle \mathbb{G}, G \;\middle|\; \begin{array}{l} [b, h_2] = 1, [a, h_1] = [a, h_3] = [b, h_1] = [b, h_3] = 1, \\ [c, h_1] = [c, h_3] = 1, [b, h_4] = [c, h_4] = [d, h_4] = 1, C \end{array} \right\rangle$$
$$= \left\langle \begin{array}{l} a, b, c, d, \\ h_1, h_2, h_3, h_4 \end{array} \;\middle|\; \begin{array}{l} [b, h_2] = 1, [a, h_1] = [a, h_3] = [b, h_1] = 1, \\ [b, h_3] = [c, h_1] = [c, h_3] = 1, \\ [b, h_4] = [c, h_4] = [d, h_4] = 1, [h_1, h_2] = 1, \\ [h_1, h_3] = 1, [h_1, h_4] = 1, [h_2, h_3] = 1, [h_3, h_4] = 1. \end{array} \right\rangle$$

The coordinate group $\mathbb{G}_{w_6}$ is the fully residually $\mathbb{G}$ partially commutative group, whose underlying commutation graph is shown on Figure 3.

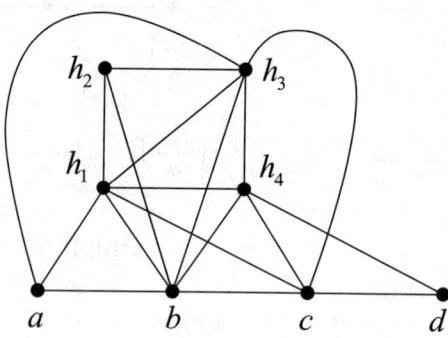

FIGURE 3. The commutation graph of $\mathbb{G}_{w_6}$.

Bibliography

[Ad75]   S. I. Adian, *Burnside problem and identities in groups*, Science, Moscow, 1975.

[Al07]   E. Alibegović, *Makanin-Razborov diagrams for limit groups*, Geom. Topol. **11** (2007) pp. 643-666. MR2302499 (2008k:57001)

[Ap68]   K. I. Appel, *One-variable equations in free groups*, Proc. Amer. Math. Soc. **19** (1968), pp. 912-918. MR0232826 (38:1149)

[Bass91]   H. Bass, *Group actions on non-Archimedean trees*, Arboreal group theory, 1991, pp. 69-130. MR1105330 (93d:57003)

[B74]   G. Baumslag, *Reviews on infinite groups as printed in Mathematical Reviews 1940 through 1970, volumes 1-40 inclusive, Part 1*, American Mathematical Society, 1974. MR0349818 (50:2311)

[BMR99]   G. Baumslag, A. G. Myasnikov, V. N. Remeslennikov, *Algebraic geometry over groups I. Algebraic sets and Ideal Theory*, J. Algebra **219** (1999), pp. 16-79. MR1707663 (2000j:14003)

[BMR02]   G. Baumslag, A. G. Myasnikov, V. N. Remeslennikov, *Discriminating Completions of Hyperbolic Groups*, Geom. Dedicata **92** (2002), pp. 115-143. MR1934015 (2003i:20073)

[BB97]   M. Bestvina, N. Brady, *Morse theory and finiteness properties of groups*, Invent. Math. **129** (1997), pp. 445-470. MR1465330 (98i:20039)

[BF95]   M. Bestvina, M. Feighn, *Stable actions of groups on real trees*, Invent. Math., **121** (1995), no. 2, pp. 287-321. MR1346208 (96h:20056)

[Bl99]   P. E. Black, ed. *Algorithms and Theory of Computation Handbook*, CRC Press LLC, 1999, "divide and marriage before conquest", in Dictionary of Algorithms and Data Structures, U.S. National Institute of Standards and Technology, Available from: http://www.nist.gov/dads/HTML/dividemarrig.html MR1797168 (2001f:68001)

[BGM06]   D. Bormotov, R. Gilman, A. Myasnikov, *Solving One-Variable Equations in Free Groups*, J. Group Theory **12** (2009), pp. 317-330 MR2502222

[Br77]   R. Bryant, *The verbal topology of a group*, J. Algebra, **48** (1977), pp. 340-346. MR0453878 (56:12131)

[Bul70]   V. K. Bulitko, *Equations and inequalities in a free group and a free semigroup.* (Russian), Tul. Gos. Ped. Inst. Učen. Zap. Mat. Kaf. no. 2 Geometr. i Algebra (1970), pp. 242-252. MR0393235 (52:14045)

[CK07]   M. Casals-Ruiz, I. Kazachkov, *Elements of Algebraic Geometry and the Positive Theory of Partially Commutative Groups*, Canad. J. Math., to appear, arXiv:0710.4077v1 [math.GR]

[CK09]   M. Casals-Ruiz, I. Kazachkov, *On Systems of Equations over Free Products of Groups*, arXiv:0903.2096v1 [math.GR]

[ChKe73]   C. C. Chang, H. J. Keisler, *Model Theory*, North-Holland, London, N.Y, 1973. MR0409165 (53:12927)

[Ch98]   O. Chapuis, *On the theories of free solvable groups*, J. Pure Appl. Algebra **131** (1998), pp. 13-24. MR1632156 (2000e:20060)

[Char07]   R. Charney, *An introduction to right-angled Artin groups*, Geom. Dedicata **125** (2007), pp. 141-158. MR2322545 (2008f:20076)

[CCV07]   R. Charney, J. Crisp, K. Vogtmann, *Automorphisms of 2-dimensional right-angled Artin groups*, Geom. Topol. **11** (2007), pp. 2227-2264. MR2372847 (2009a:20059)

[CR00]   I.M.Chiswell, V.N.Remeslennikov, *Equations in Free Groups with One Variable: I*, J. Group Theory **3** (2000), pp. 445-466. MR1790341 (2001i:20054)

[Com81] L. P. Comerford, *Quadratic equations over small cancellation groups*, J. Algebra **69** (1981), no. 1, pp. 175-185. MR613867 (82f:20060)

[ComEd81] L. P. Comerford, C. C. Edmunds, *Quadratic equations over free groups and free products*, J. Algebra **68** (1981), no. 2, pp. 276-297. MR608536 (82k:20060)

[ComEd89] L. P. Comerford, C. C. Edmunds, *Solutions of equations in free groups*, Walter de Gruyter, Berlin, New York, 1989. MR981853 (90a:20067)

[CW04] J. Crisp and B. Wiest, *Embeddings of graph braid groups and surface groups in right-angled Artin groups and braid groups*, Algebr. Geom. Topol. **4** (2004), pp. 439-472. MR2077673 (2005e:20052)

[Cul81] M. Culler, *Using surfaces to solve equations in free groups*, Topology **20** (1981), no. 2, pp. 133-145. MR605653 (82c:20052)

[Dahm09] F. Dahmani, *Existential questions in (relatively) hyperbolic groups*, Israel J. Math **173** (2009), pp. 91-124. MR2570661

[DG09] F. Dahmani, V. Guirardel, *Foliations for solving equations in groups: free, virtually free, and hyperbolic groups*, J. Topol, to appear, arXiv:0901.1830v2 [math.GR]

[Day08a] M. Day, *Peak reduction and finite presentations for automorphism groups of right-angled Artin groups*, arXiv:0807.4799v1 [math.GR]. Geom. Topol. **13** (2009), 817–855. MR2470964 (2009m:20052)

[Day08b] M. Day, *Symplectic structures on right-angled Artin groups: between the mapping class group and the symplectic group*, arXiv:0807.4801v1 [math.GR]. Geom. Topol. **13** (2009), 857–899. MR2470965 (2010b:20065)

[DGH01] V. Diekert, C. Gutiérrez, C. Hagenah, *The existential theory of equations with rational constraints in free groups is PSPACE-complete*, Inform. and Comput. **202** (2005), pp. 105-140. MR2172984 (2006h:68048)

[DL04] V. Diekert, M. Lohrey, *Existential and Positive Theories of Equations in Graph Products*, Theory Comput. Syst. **37** (2004), pp. 133-156. MR2038406 (2005a:03023)

[DMM99] V. Diekert, Yu. Matiyasevich, A. Muscholl *Solving word equations modulo partial commutations*, Theoret. Comput. Sci. **224** (1999), pp. 215-235. MR1714796 (2001g:03023)

[DM06] V. Diekert, A. Muscholl *Solvability of Equations in Graph Groups is Decidable*, Internat. J. Algebra Comput. **16** (2006), pp. 1047-1069. MR2286422 (2007k:03021)

[DK93] G. Duchamp, D. Krob, *Partially Commutative Magnus Transformations*, Internat. J. Algebra Comput. **3** (1993), 15-41. MR1214003 (94e:20041)

[Dun07] A. J. Duncan, *Exponential genus problems in one-relator products of groups*, Mem. Amer. Math. Soc. no. 873 (2007). MR2294597 (2008m:20047)

[DKR06] A. J. Duncan, I. V. Kazachkov, V. N. Remeslennikov, *Centraliser dimension and universal classes of groups*, SEMR **3** (2006), pp. 197-215. MR2276020 (2007k:20002)

[DKR07] A. Duncan, I. Kazachkov, V. Remeslennikov, *Parabolic and quasiparabolic subgroups of free partially commutative groups*, J. Algebra, **318** (2007), pp. 918-932. MR2371978 (2009j:20057)

[DKR08] A. J. Duncan, I. V. Kazachkov, V. N. Remeslennikov, *Automorphisms of Partially Commutative Groups I: Linear Subgroups*, Groups Geom. Dyn., to appear, arXiv:0803.2213v1 [math.GR]

[Edm75] C. C. Edmunds, *On the endomorphism problem for free groups*, Comm. Algebra **3** (1975), pp. 1-20. MR0369530 (51:5763)

[Edm77] C. C. Edmunds, *On the endomorphism problem for free groups II*, Proc. London Math. Soc. **38** (1979), pp. 153-168. MR520977 (80d:20025)

[Eis99] D. Eisenbud, *Commutative Algebra with a View Toward Algebraic Geometry*, Springer, 1999. MR1322960 (97a:13001)

[EKR05] E. S. Esyp, I. V. Kazachkov and V. N. Remeslennikov, *Divisibility Theory and Complexity of Algorithms for Free Partially Commutative Groups*, In: Groups, Languages, Algorithms. Contemp. Math. 378 (2005), 319-348. MR2159318 (2006f:20038)

[Ful04] S. Fulthorp, *Genus n Forms over Hyperbolic Groups*, M.Sc. Thesis, University of Newcastle upon Tyne, 2004.

[GLP94] D. Gaboriau, G. Levitt, F. Paulin, *Pseudogroups of isometries of R and Rips' theorem on free actions on R-trees*, Israel J. Math., **87** (1994), 403-428. MR1286836 (95e:20042)

[Gr90] E. R. Green *Graph products of groups*, Ph.D. Thesis, University of Leeds, 1990.

BIBLIOGRAPHY

[GK89]   R. I. Grigorchuk, P. F. Kurchanov, *On quadratic equations in free groups*, Proceedings of the International Conference on Algebra, Part 1 (Novosibirsk, 1989), pp. 159-171; Contemp. Math. 131, Part 1, Amer. Math. Soc., Providence, RI, 1992. MR1175769 (94m:20074)

[GrL92] R. I. Grigorchuk, I. G. Lysënok, *A description of solutions of quadratic equations in hyperbolic groups*, Internat. J. Algebra Comput. **2** (1992), no. 3, pp. 237-274. MR1189234 (94d:20033)

[GPR07] M. Gutierrez, A. Piggott, K. Ruane, *On the automorphisms of a graph product of abelian groups*, arXiv:0710.2573v1 [math.GR]

[Gr05]   D. Groves, *Limit groups for relatively hyperbolic groups, II: Makanin-Razborov diagrams*, Geom. Topol. **9** (2005), pp. 2319-2358. MR2209374 (2007e:20086)

[Gr08]   D. Groves, *Limit groups for relatively hyperbolic groups, I: The basic tools*, arXiv:0412492v2 [math.GR]. Algebr. Geom. Topol. **9** (2009), 1423–1466. MR2530123

[Guba86] V. Guba, *Equivalence of infinite systems of equations in free groups and semigroups to finite subsystems*, Mat. Zametki, **40** (1986), pp. 321-324. MR869922 (88d:20060)

[HW08]   F. Haglund, D. Wise, *Special Cube Complexes*, Geom. Funct. Anal. **17** (2008), pp. 1551-1620. MR2377497 (2009a:20061)

[HNN49] G. Higman, B. Neumann, H. Neumann, *Embedding Theorems for Groups*, J. London Math. Soc. s1-24 (1949), no.4, pp. 247-254. MR0032641 (11:322d)

[Hm71]   Yu. I. Hmelevskiĭ, *Systems of equations in a free group. I*, (Russian), Izv. Akad. Nauk. Ser. Mat., **35** (1971), no. 6, pp. 1237-1268; translation in Math. USSR Izv. **5** (1971), pp. 1245-1276. MR0313395 (47:1949)

[Hm72]   Yu. I. Hmelevskiĭ, *Systems of equations in a free group. II*, (Russian), Izv. Akad. Nauk. Ser. Mat., **36** (1972), no. 1, pp. 110-179, translation in Math. USSR Izv. **6** (1972), pp. 109-180. MR0313395 (47:1949)

[Hum94] S. Humphries, *On representations of Artin groups and the Tits conjecture*, J. Algebra **169** (1994), pp. 847-862. MR1302120 (95k:20057)

[JS09]   E. Jaligot, Z. Sela, *Makanin-Razborov Diagrams over Free Products*, Illinois J. Math., to appear, arXiv:0902.2493v2 [math.GR]

[Kaz07] I. Kazachkov, *Algebraic Geometry over Lie Algebras*, Surveys in Contemporary Mathematics, LMS Lecture Notes, **347** (2007), pp. 34-81. MR2388490 (2009g:17035)

[KhLMT08] O. Kharlampovich, I. G. Lysënok, A. G. Myasnikov, N. W. M. Touikan, *Quadratic equations over free groups are NP-complete*, Theory Comput. Syst. **47** (2010), pp. 250-258.

[KhM98a] O. Kharlampovich, A. Myasnikov, *Irreducible affine varieties over a free group I: Irreducibility of quadratic equations and Nullstellensatz*, J. Algebra, **200** (1998), pp. 472-516. MR1610660 (2000b:20032a)

[KhM98b] O. Kharlampovich, A. Myasnikov, *Irreducible affine varieties over a free group II: Systems in triangular quasi-quadratic form and description of residually free groups*, J. of Algebra, **200** (1998), pp. 517-570. MR1610664 (2000b:20032b)

[KhM05a] O. Kharlampovich, A. Myasnikov, *Implicit function theorem over free groups*, J. Algebra, **290** (2005), pp. 1-203. MR2154989 (2007b:20047)

[KhM05b] O. Kharlampovich, A. Myasnikov, *Effective JSJ Decompositions*, In: Groups, Languages, Algorithms. Contemp. Math. 378 (2005), pp. 87-212. MR2159316 (2006m:20045)

[KhM06] O. Kharlampovich, A. Miasnikov, *Elementary theory of free non-abelian groups*, J.Algebra, **302** (2006), pp. 451-552. MR2293770 (2008e:20033)

[LS08]   M. Lohrey, G. Sénizergues, *Equations in HNN extensions*, preprint, http://inf.informatik.uni-stuttgart.de/fmi/ti/personen/Lohrey/HNN.html

[Lor63] A. A. Lorents, *The solution of systems of equations in one unknown in free groups*, Dokl. Akad. Nauk **148** (1963), pp. 1253-1256. MR0183809 (32:1285)

[Lor68] A. A. Lorents, *Representations of sets of solutions of systems of equations with one unknown in a free group*, Dokl. Akad. Nauk **178** (1968), pp. 290-292. MR0225861 (37:1452)

[Lyn60] R. C. Lyndon, *Equations in free groups*, Trans. Amer. Math. Soc. **96** (1960), pp.445-457. MR0151503 (27:1488)

[Lyn80] R. Lyndon, *Equations in groups*, Bull. Braz. Math. Soc. (N.S.) **11** (1980), pp. 79-20 MR607019 (82j:20070)

BIBLIOGRAPHY

[LS77] R. C. Lyndon, P. E. Schupp, *Combinatorial group theory*, Springer, 1977. MR0577064 (58:28182)

[Lys88] I. G. Lysënok, *Solutions of quadratic equations in groups with the small cancellation condition*, (Russian) Mat. Zametki **43** (1988), no. 5, pp. 577-592; translation in Math. Notes **43** (1988), no. 5-6, pp. 333-341 MR954341 (90b:20029)

[Mak77] G. S. Makanin, The problem of solvability of equations in a free semigroup, Mat. Sb. (N.S.), 1977, 103(145):2(6), pp. 147-236. MR0470107 (57:9874)

[Mak82] G. S. Makanin, *Equations in a free group* (Russian), Izv. Akad. Nauk SSSR, Ser. Mat., **46** (1982), pp. 1199-1273, transl. in Math. USSR Izv., **21** (1983) MR682490 (84m:20040)

[Mak84] G. S. Makanin, *Decidability of the universal and positive theories of a free group* (Russian), Izv. Akad. Nauk SSSR, Ser. Mat., **48** (1985), pp. 735-749, transl. in Math. USSR Izv., **25** (1985). MR755956 (86c:03009)

[Mal62] A. I. Malcev. *On the equation $zxyx^{-1}y^{-1}z^{-1} = aba^{-1}b^{-1}$ in a free group*, Algebra i Logika Sem. **1** (1962), no. 5, pp. 45-50. MR0153726 (27:3687)

[Mat97] Yu. Matiyasevich, *Some decision problems for traces*, In S. Adian and A. Nerode, editors, Proceedings of the 4th International Symposium on Logical Foundations of Computer Science (LFCS'97), Yaroslavl, Russia, July 6-12, 1997, number 1234 in Lecture Notes in Computer Science, pp. 248-257, Berlin, Heidelberg, 1997, Springer-Verlag, Invited lecture. MR1611426

[McC75] J. McCool, *Some finitely presented subgroups of the automorphism group of a free group*, J. Algebra **35** (1975), pp. 205-213. MR0396764 (53:624)

[Mer66] Yu. I. Merzlyakov. *Positive formulae on free groups*. Algebra i Logika **5** (1966), pp. 25-42. MR0222149 (36:5201)

[MR96] A. G. Myasnikov, V. N. Remeslennikov, *Exponential groups 2: extension of centralizers and tensor completion of CSA-groups*, Internat. J. Algebra Comput., **6** (1996), pp. 687-711. MR1421886 (97j:20039)

[MR00] A. G. Myasnikov, V. N. Remeslennikov, *Algebraic geometry over groups II. Logical Foundations.* J. Algebra **234** (2000), pp. 225-276. MR1799485 (2001i:14001)

[Niel] J. Nielsen, *Die Isomorphismen der allgemeinen, unendlichen Gruppe mit zwei Erzeugenden*, Math. Ann. **78** (1918), pp. 385-397. MR1511907

[Oj83] Yu. I. Ozhigov, *Equations in two variables over a free group*, Dokl. Akad. Nauk, **268** (1983), no. 4, pp. 809-813. MR693212 (84j:20033)

[Ol89] A. Yu. Ol'shanskii, *Diagrams of homomorphisms of surface groups*, (Russian) Sibirsk. Mat. Zh. **30** (1989), no. 6, pp. 150-171; translation in Siberian Math. J. **30** (1989), no. 6, pp. 961-979. MR1043443 (91e:20028)

[Pl99a] W. Plandowski, *Satisfiability of word equations with constants is in NEXPTIME*, In Theory of Computing (STOC 99), pp. 721-725. ACM 49. Press. MR1798096 (2001h:68059)

[Pl99b] W. Plandowski, *Satisfiability of word equations with constants is in PSPACE*, In Foundations of Computer Science (FOCS 99), pp. 495-500. IEEE Computer Society Press. MR1917589

[Raz84] A. A. Razborov, *An equation in a free group whose set of solutions does not allow a representation as a superposition of a finite number of parametric functions*, In: Proceedings of the 9th All-Union Symposium on Group Theory, Moscow, 1984, p. 54.

[Raz85] A. A. Razborov, *On systems of equations in a free group*, Math. USSR, Izvestiya, **25** (1985), pp. 115-162. MR755958 (86c:20033)

[Raz87] A. A. Razborov, *On systems of equations in a free group*, Ph.D. thesis, Steklov Math. Institute, Moscow, 1987.

[Raz95] A. A. Razborov *On systems of equations in free groups*, in Combinatorial and geometric group theory. Edinburgh 1993, pp. 269-283, Cambridge University Press, 1995. MR1320290 (96c:20039)

[Rem89] V. N. Remeslennikov, *∃-free groups*, Siberian Math. J., **30** (1989), no. 6, pp. 153-157. MR1043446 (91f:03077)

[Rep83] N. N. Repin, *Equations with one unknown in nilpotent groups*, (Russian) Mat. Zametki **34** (1983), no. 2, pp. 201-206, English translation in Math. Notes **34** (1983), no. 1-2, pp. 582-585. MR719474 (85g:20045)

[Rep84] N. N. Repin, *Solvability of equations with one indeterminate in nilpotent groups* (Russian) Izv. Akad. Nauk SSSR Ser. Mat. **48** (1984), no. 6, pp. 1295-1313. MR772117 (86d:20040)

[RS95] E. Rips, Z. Sela, *Canonical representatives and equations in hyperbolic groups*, Invent. Math. **120**, pp. 489-512. MR1334482 (96c:20053)

[RS97] E. Rips, Z. Sela, *Cyclic splittings of finitely presented groups and the canonical JSJ decomposition*, Ann. of Math. (2) **146** (1997), pp. 53-109. MR1469317 (98m:20044)

[Rom77] V. A. Roman'kov, *Unsolvability of the endomorphic reducibility problem in free nilpotent groups and in free rings* (Russian) Algebra i Logika **16** (1977), no. 4, pp. 457-471; English translation in: Algebra and Logic, **16** (1977), no. 4, pp. 310-320. MR516297 (81b:20027)

[Rom79a] V. A. Roman'kov, *Universal theory of nilpotent groups*, (Russian), Mat. Zametki **25** (1979), no. 4, pp. 487-495. MR534291 (80j:03058)

[Rom79b] V. A. Roman'kov, *On equations in free metabelian groups*, Siberian Math. J. **20** (1979), no. 3, pp. 671-673. MR537377 (80k:20040)

[Sch90] K. Schulz, *Makanin's Algorithm for Word Equations - Two Improvements and a Generalization*, Proceedings of the First International Workshop on Word Equations and Related Topics, pp. 85-150, 1990. MR1232032 (94k:20119)

[Sch69] P. E. Schupp, *On the substitution problem for free groups*, Proc. Amer. Math. Soc. **23** (1969), pp. 421-24. MR0245657 (39:6963)

[Sela97] Z. Sela, *Structure and rigidity in (Gromov) hyperbolic groups and discrete groups in rank 1 Lie groups, II*, Geom. Funct. Anal. **7** (1997), 561-593. MR1466338 (98j:20044)

[Sela01] Z. Sela, *Diophantine geometry over groups I: Makanin-Razborov diagrams*, Publ. Math. Inst. Hautes Études Sci. **93** (2001), pp. 31-105. MR1863735 (2002h:20061)

[Sela03] Z. Sela, *Diophantine geometry over groups II: Completions, closures and formal solutions*, Israel J. Math. **134** (2003), pp. 173-254. MR1972179 (2004g:20061)

[Sela06] Z. Sela, *Diophantine geometry over groups VI: The elementary theory of a free group*, Geom. Funct. Anal. **16** (2006), 707-730. MR2238945 (2007j:20063)

[Sela02] Z. Sela, *Diophantine geometry over groups VII: The elementary theory of a hyperbolic group*, Proc. Lond. Math. Soc. (3), **99** (2009), pp. 217-273. MR2520356

[Ser89] H. Servatius, *Automorphisms of Graph Groups*, J. Algebra, **126** (1989), no. 1, pp. 34-60. MR1023285 (90m:20043)

[Sh05] S. L. Shestakov, *The equation $[x,y] = g$ in partially commutative groups*, Siberian Math. J., **46** (2005), pp. 364-372; translation in Siberian Math. J., **46** (2005), no. 2, pp. 364-372. MR2141291 (2007c:20082)

[Sh06] S. L. Shestakov, *The equation $x^2 y^2 = g$ in partially commutative groups*, Sibirsk. Mat. Zh., **47** (2006), no.2, pp. 463-472; translation in Siberian Math. J., **47** (2006), no. 2, pp. 383-390. MR2227990 (2007f:20064)

[T08] N. Touikan, *On the coordinate groups of irreducible systems of equations in two variables over free groups*, arXiv:0810.1509v1 [math.GR]

[VW94] L. Van Wyk, *Graph groups are biautomatic*, J. Pure Appl. Algebra **94** (1994), no.3, pp. 341-352. MR1285550 (95g:20041)

[Vd97] A. Vdovina, *Products of Commutators in Free Products*, Inter. Jour. of Alg. and Computation, **7** (1997), no. 4, pp. 471-485. MR1459623 (98k:20059)

[Wic62] M. J. Wicks, *Commutators in Free Products*, J. London Math. Soc. **37** (1962), pp. 433-444. MR0142610 (26:179)

Index

algebraic set, 14
automorphism
  completely induced, 100
  dual, 100
  invariant with respect to the kernel, 101
  invariant with respect to the kernel in the sense of Razborov, 101
  invariant with respect to the non-quadratic part, 102
  invariant with respect to the non-quadratic part in the sense of Razborov, 103
  tame, 101

base
  active, 51
  belongs to a section, 31
  carrier, 60
  constant, 27
  contained in another base, 30
  contains an item, 30
  dual, 27
  eliminable, 57
  intersects another base, 30
  leading, 60
  linear, 30
  non-active, 51
  of a generalised equation, 27
  overlaps with another base, 30
  quadratic, 102
  quadratic-coefficient, 102
  transfer, 60
  variable, 27
block, 19
boundary
  μ-tied, 30
  closed, 30
  free, 30
  intersects a base, 30
  of a generalised equation, 27
  open, 30
  tied, 30
  touches a base, 30
boundary connection, 27

cancellation divisor, 71
cancellation matrix, 73
canonical homomorphism of coordinate groups of generalised equations, 63
canonical parabolic subgroup, 18
carrier, 60
clan, 22
  thick, 22
  thin, 22
complexity of the generalised equation, 63
coordinate group
  of a system of equations, 15
  of an algebraic set, 15
  of the (constrained) generalised equation, 34
cyclically reduced element, 18

dependence graph, 22
G-discriminated group, 17
divisor of an element, 18

elimination of a base, 58
elimination process for a generalised equation, 58
equation of a generalised equation
  basic, 27
  boundary, 28
  coefficient, 28
equation over a group, 14
G-equationally Noetherian group, 15
equivalence
  of generalised equations, 29
  of systems of equations, 15
excess of the solution, 61
exponent
  of (P-)periodicity, 77
  of a word, 77

formula
  atomic, 16
  universal, 16
G-fully residually G group, 17

generalised equation, 29
  combinatorial, 27

constrained, 29
formally consistent, 31
periodised, 81
quadratic, 61
regular with respect to a periodic structure, 86
singular with respect to a periodic structure, 86
strongly singular with respect to a periodic structure, 85
geodesic, 18
global position, 23
graph
commutation, of a partially commutative group, 18
non-commutation, of a partially commutative group, 19
of a periodic structure, 80
graph group, 17
graph product of groups, 22
group
recursive, 101
G-group, 13
finitely generated, 14

height
of a tree, 13
of a vertex, 13
G-homomorphism, 13

irreducible
algebraic set, 15
component, 15
isomorphism of algebraic sets, 15
item
active, 51
belongs to a base, 30
belongs to a section, 31
belongs to the kernel, 58
constant, 30
covered γ_i times, 30
free, 30
linear, 30
non-active, 51
of a generalised equation, 27
quadratic, 30

kernel of a generalised equation, 57

leaf
final of the tree T, 111
final of the tree T_0, 119
final of the tree T_{dec}, 127
final of the tree T_{ext}, 129
final of the tree T_{sol}, 135
left-divide, 18
length, 18
of a solution of the generalised equation, 30

letter, 18

median position, 23
morphism of algebraic sets, 15

normal form, 23
DM-, 23
lexicographical, 18

occurrence, 18

pair of matched bases, 30
part of a generalised equation
active, 51
non-active, 51
non-quadratic, 102
quadratic, 102
partially commutative group, 17
$\mathbb{G}$-partition table, 32
path
μ-reducing, 108
$\mathfrak{p}(H)$, 99, 110
defined by a solution in the tree $T(\Omega)$, 64
prohibited of type 12, 104
prohibited of type 15, 108
prohibited of type 7-10, 103
period, 77
(P-)periodic word, 77
periodic structure, 78
connected, 80
regular, 86
singular, 86
strongly singular
of type (a), 86
of type (b), 86

quasi identity, 16

radical
of a set, 14
of a system, 14
G-residually G group, 17
right-angled Artin group, 17
right-divide, 18

section, 31
(P-)periodic, 78
active, 51
closed, 31
constant, 51
non-active, 51
variable, 51
sentence, 16
G-separated group, 17
solution
associated to the vertex, 133
of a constrained generalised equation, 30
of a generalised equation
minimal with respect to the group of automorphisms, 74

over $\mathbb{F}$, 30
over $\mathbb{T}$, 29
periodic with respect to a period, 77
of a system of equations, 14
source point, 23
standard form of a generalised equation, 51
G-subgroup, 14
system of equations over a group, 14

target point, 23
terminal vertex, 128
transformation of a generalised equation
derived, 56
elementary, 51
entire, 60
type
of a generalised equation, 68
of a vertex, 68

variable of a generalised equation, 27

Glossary of notation

$A(\Omega_v)$: the automorphism group associated to a vertex v of the trees $T_{\mathrm{dec}}(\Omega)$, $T_{\mathrm{ext}}(\Omega)$ and $T_{\mathrm{sol}}(\Omega)$, 126, 129, 135

$A\Sigma$: the active part of a generalised equation, 51

$\mathfrak{A}(\Omega)$: finitely generated group of automorphisms of $\mathbb{G}_{R(\Omega^*)}$ associated with a periodic structure on Ω, 86

$\mathrm{Aut}(\Omega)$: recursive group of automorphisms associated to the root of $T_0(\Omega)$, 109

$\mathbb{A}(W)$: the subgroup of $\mathbb{G}$ generated by all letters that do not occur in $\overline{w}$ and commute with w for every $w \in W$, 18

$\mathbb{A}(w)$: the subgroup of $\mathbb{G}$ generated by all letters that do not occur in $\overline{w}$ and commute with w, 18

α: function from the set $\mathcal{BS}$ of bases to the set $\mathcal{BD}$ of boundaries of a generalised equation that determines the left-most boundary of a base, 27

$\mathrm{alph}(W)$: the set of letters that occur in a word $w \in W$, 18

$\mathrm{alph}(w)$: the set of letters that occur in w, 18

$\mathcal{A}$: a finite alphabet, the generating set of $\mathbb{G}$, 18

$\mathcal{BC}$: the set of boundary connections of a generalised equation, 27

$\mathcal{BD}$: the set of boundaries of a generalised equation, 27

$\mathcal{BS}$: the set of bases of a generalised equation, 27

β: function from the set $\mathcal{BS}$ of bases to the set $\mathcal{BD}$ of boundaries of a generalised equation that determines the right-most boundary of a base, 27

$C\Sigma$: the set of all constant sections of a generalised equation, 50

$C^{(1)}$: "short" cycles in Γ, 82

$C^{(2)}$: "free" cycles in Γ, 82

$[A, B]$: the set of commutators of elements of sets A and B, 18

$[A, B] = 1$: elements of sets A and B commute pairwise, 18

comp, $\mathrm{comp}(\Omega)$: complexity of a generalised equation, 63

$\mathrm{CD}(u, v)$: the cancellation divisor of u and v, 71

D 1: derived transformation D 1, 56
D 2: derived transformation D 2, 57
D 3: derived transformation D 3, 57
D 4: derived transformation D 4, 57
D 5: derived transformation D 5, 60
D 6: derived transformation D 6, 61

Δ: involution on the set of variable bases of a generalised equation that determines the dual of a base, 27

$\delta(l)$: decomposition of the period defined by the boundary l, 79

$d_1(H)$: length of the "quadratic part of the solution" H, 113

$d_2(H)$: length of the "quadratic-coefficient part of the solution" H, 113

$d_{A\Sigma}(H)$: length of the active part of the solution, $d_{A\Sigma}(H) = \sum_{i=1}^{\rho_A-1} |H_i|$, 61

ET 1: elementary transformation ET 1, 52
ET 2: elementary transformation ET 2, 52
ET 3: elementary transformation ET 3, 52
ET 4: elementary transformation ET 4, 53
ET 5: elementary transformation ET 5, 54

$\exp(w)$: exponent of periodicity of the word w, 77

e: a function that assigns a pair $(\Omega_{v'}, H^{(v')})$ to the pair $(\Omega_v, H^{(v)})$, $\mathrm{tp}(v) \neq 1, 2$, 110

e': a function that assigns a pair $(\Omega_{v'}, H^{(v')})$ to the pair $(\Omega_v, H^{(v)})$, $\mathrm{tp}(v) = 15$, 110

ε: function from the set of variable bases of a generalised equation to $\{-1, 1\}$ that determines orientation of the base, 27

$\mathbb{F}$, $\mathbb{F}(\mathcal{A}^{\pm 1})$: free monoid on the alphabet $\mathcal{A} \cup \mathcal{A}^{-1}$, 22

$\mathbb{G}_{R(S)}$: coordinate group of the system S over $\mathbb{G}$, 15

$\mathcal{GE}'(S)$: the set of all constrained generalised equations $\Omega'_\mathcal{T}$ over $\mathbb{T}$ constructed by $\mathbb{G}$-partition tables $\mathcal{T}$ for the system $S(X, \mathcal{A})$, 34

$\mathcal{GE}(\Omega')$: the set of all constrained generalised equations Ω_T over $\mathbb{F}$ constructed for the constrained generalised equation Ω' over $\mathbb{T}$, 39

$\mathbb{G}$: partially commutative group, 18

$\mathbb{G}(\mathcal{G})$: partially commutative group with underlying commutation graph $\mathcal{G}$, 18

$\mathbb{G}_{R(\Omega^*)}$: coordinate group of the generalised equation Ω, 34

$\mathbb{G}_{w_i}$: fully residually $\mathbb{G}$ partially commutative group associated to the leaf of the solution tree $T_{\mathrm{sol}}(\Omega)$, 133

Γ: graph of a periodic structure, 80

Γ_0: the subgraph of Γ all of whose edges are "short", $\Gamma_0 = (V(\Gamma), \mathrm{Sh}(\Gamma))$, 81

F: a non-oriented graph that encodes all canonical parabolic subgroups of the group $\mathbb{G}$ and the $\leftrightarrows$-commutativity relation between them, 132

$\gamma(h_i), \gamma_i$: the number of bases which contain h_i, 30

$\gamma_i(\omega)$: the number of bases from ω_1 that contain h_i, 115

$\beth$: a non-oriented graph that describes the canonical parabolic subgroups associated to the solution of a generalised equation, 133

$\mathcal{G}$: commutation graph of a partially commutative group, 17

$H(\mu)$: if H is a solution of a generalised equation, $H(h(\mu)) = H_{\alpha(\mu)} \cdots H_{\beta(\mu)-1}$, 31

$H[i,j]$, $H(\sigma)$: if H is a solution of a generalised equation, $H(\sigma) = H_i \cdots H_{j-1}$, 31

$\mathrm{Hom}_G(H, K)$: set of G-homomorphisms from H to K, 13

Glossary of notation

$|H|$: length of a solution of a (constrained) generalised equation, $|H| = \sum_{i=1}^{\rho} |H_i|$, 30

$h(\mathrm{Sh})$: labels of the "short" edges of Γ, $h(\mathrm{Sh}) = \{h(e) \mid e \in \mathrm{Sh}\}$, 81

$h(\mu)$: the product of items $h_{\alpha(\mu)} \ldots h_{\beta(\mu)-1}$, 30

$h[i, j]$: the product of items $h_i \ldots h_{j-1}$, 30

$\mathrm{Ker}(\Omega)$, $\mathrm{Ker}(\Upsilon)$: kernel of a (constrained) generalised equation, 57

$\mathcal{L}$: first-order language of groups, 15

$\mathcal{L}_A$: first-order language of groups enriched by constants from A, 15

$\mathcal{L}_G$: first-order language of groups enriched by constants from G, 15

$\mathbb{M}$: free or partially commutative monoid, $\mathbb{M} = \mathbb{F}(\mathcal{A}^{\pm 1})$ or $\mathbb{M} = \mathbb{T}(\mathcal{A}^{\pm 1})$, 27

$NA\Sigma$: the non-active part of a generalised equation, 51

$\mathrm{ncl}\langle S \rangle$: normal closure of S, 14

$n(\sigma)$: the number of bases in the section σ, 63

n_A, $n_A(\Omega)$: the number of bases in the active sections of a generalised equation, 63

(Ω, H): H is a solution of the constrained generalised equation Ω, 30

Ω: constrained generalised equation, 30

$\langle \Upsilon, \Re_\Upsilon \rangle$: constrained generalised equation, 29

ω_1: the set of all variable bases ν for which either ν or $\Delta(\nu)$ belongs to the active part of a generalised equation, 61

$\widetilde{\Omega}$, $\widetilde{\Upsilon}$: the generalised equation obtained from Ω (or Υ) by D 3, 57

Ω^*: the generalised equation Ω treated as a system of equations over the partially commutative group $\mathbb{G}$, 34

$\mathcal{P}$: non-empty set of variables, variable bases, and closed sections that belong to the periodic structure $\langle \mathcal{P}, R \rangle$, 78

Π_v: a non-oriented graph that encodes coefficient equations of Ω_v and the relation $\Re_{\Upsilon_v}$, 132

$\langle \mathcal{P}, R \rangle$: periodic structure, 78

π_U: homomorphism defined by the solution U, 14

$\psi_{A\Sigma}(H)$: excess of the solution, $\psi_{A\Sigma}(H) = \sum_{\mu \in \omega_1} |H(\mu)| - 2d_{A\Sigma}(H)$, 61

$\mathcal{P}(H, P)$: periodic structure associated to a P-periodic solution H, 79

$R(S)$: radical of the system S, 14

$R(Y)$: radical of a set Y, 14

$\Re(x)$: set of elements related with x, 13

ρ, ρ_Υ, ρ_Ω: the number of items of a generalised equation, 27

ρ_A: the boundary between active and non-active parts of a generalised equation, 51

ρ'_v: the number of free variables that belong to the non-active sections of Ω_v, 107

$\mathfrak{r}$: the cardinality of $\mathcal{A}$, 18

S_Υ: system of equations associated to the generealised equation, 27

$[i,j]$: section $\{h_i, \ldots, h_{j-1}\}$ of a generalised equation, 30

Sh, Sh(Γ): the set of "short" edges of the graph Γ of a periodic structure, 81

Σ, $\Sigma(\Omega)$: the set of all closed sections of a generalised equation Ω, 50

$\mathbf{s}(\Omega_v)$: a function that bounds the length of a minimal solution, 108

$\sigma(\mu)$: the section $[\alpha(\mu), \beta(\mu)]$, 30

$T(\Omega)$: the infinite, locally finite tree of the process, 62

$T_0(\Omega)$: the finite subtree of $T(\Omega)$ that does not contain prohibited paths, 103, 109

$T_{\text{dec}}(\Omega)$: the decomposition tree of Ω, 121

$T_{\text{ext}}(\Omega)$: the extension tree of Ω, 127

T_{sol}: the solution tree of a generalised equation, 135

$\mathbb{T}$, $\mathbb{T}(\mathcal{A}^{\pm 1})$: partially commutative monoid monoid on the alphabet $\mathcal{A} \cup \mathcal{A}^{-1}$ corresponding to the group $\mathbb{G}$, 22

τ_v, $\tau(\Omega_v)$: function of the generalised equation, 107

$\text{tp}(v)$: type of the vertex v of the tree T, T_0, T_{dec}, T_{ext} or T_{sol}, 68

(Υ, H): H is a solution of the generalised equation Υ, 30

Υ: generalised equation, 30

$\Upsilon(S)$: combinatorial generalised equation associated to a system of equations over a monoid, 28

$\bar{u}$: a subset of the set $\bar{x}$ of generators of the coordinate group of a periodised generalised equation, 83

Υ^*: generalised equation Υ treated as a system of equations over the partially commutative group $\mathbb{G}$, 34

$V\Sigma$: the set of all variable sections of a generalised equation, 50

$V_G(S)$: algebraic set defined by S over G, 14

$\mathfrak{V}(\Omega_v)$: a recursive group of automorphisms of $\mathbb{G}_{R(\Omega_v^*)}$ that we associate to a vertex v of the tree $T(\Omega)$, 100

$W(H)$: for any word $W(h)$ in $\mathbb{G}[h]$ and any solution H of a generalised equation, $W(H) = H(W(h))$, 31

$|w|$: length of the word $w \in \mathbb{G}$, 18

$\overline{w}$: a geodesic of a word w, $w \in \mathbb{G}$, 18

$\bar{x}$: a set of generators of the coordinate group of a periodised generalised equation, 83

ξ: the number of open boundaries in the active sections of a generalised equation, 63

$\bar{z}$: a subset of the set $\bar{x}$ of generators for the coordinate group of a periodised generalised equation, 83

'$<_{\mathfrak{C}(\Omega)}$': reflexive, transitive relation on the set of solutions of a generalised equation, 71

'$H^{(i)} <_{\text{Aut}(\Omega)} H^{(i')}$': if $H^{(i)}$ is a solution of Ω_{v_i} and $H^{(i')}$ is a solution of $\Omega_{v_{i'}}$, $v_i, v_{i'} \in T_0(\Omega)$, then $H^{(i)} <_{\text{Aut}(\Omega)} H^{(i')}$ if and only if $H^{(i)} <_{\pi(v_0, v_{i'})^{-1} \text{Aut}(\Omega) \pi(v_0, v_i)} H^{(i')}$, 109

'$\doteq$': graphical equality of words, 18

'$\rightleftarrows$': disjoint commutation of elements or sets, 18
'$u \circ v$': no cancellation in the product of u and v, 18

Editorial Information

To be published in the *Memoirs*, a paper must be correct, new, nontrivial, and significant. Further, it must be well written and of interest to a substantial number of mathematicians. Piecemeal results, such as an inconclusive step toward an unproved major theorem or a minor variation on a known result, are in general not acceptable for publication.

Papers appearing in *Memoirs* are generally at least 80 and not more than 200 published pages in length. Papers less than 80 or more than 200 published pages require the approval of the Managing Editor of the Transactions/Memoirs Editorial Board. Published pages are the same size as those generated in the style files provided for $\mathcal{AMS}$-LaTeX or $\mathcal{AMS}$-TeX.

Information on the backlog for this journal can be found on the AMS website starting from http://www.ams.org/memo.

A Consent to Publish and Copyright Agreement is required before a paper will be published in the *Memoirs*. After a paper is accepted for publication, the Providence office will send a Consent to Publish and Copyright Agreement to all authors of the paper. By submitting a paper to the *Memoirs*, authors certify that the results have not been submitted to nor are they under consideration for publication by another journal, conference proceedings, or similar publication.

Information for Authors

Memoirs is an author-prepared publication. Once formatted for print and on-line publication, articles will be published as is with the addition of AMS-prepared frontmatter and backmatter. Articles are not copyedited; however, confirmation copy will be sent to the authors.

Initial submission. The AMS uses Centralized Manuscript Processing for initial submissions. Authors should submit a PDF file using the Initial Manuscript Submission form found at www.ams.org/submission/memo, or send one copy of the manuscript to the following address: Centralized Manuscript Processing, MEMOIRS OF THE AMS, 201 Charles Street, Providence, RI 02904-2294 USA. If a paper copy is being forwarded to the AMS, indicate that it is for *Memoirs* and include the name of the corresponding author, contact information such as email address or mailing address, and the name of an appropriate Editor to review the paper (see the list of Editors below).

The paper must contain a *descriptive title* and an *abstract* that summarizes the article in language suitable for workers in the general field (algebra, analysis, etc.). The *descriptive title* should be short, but informative; useless or vague phrases such as "some remarks about" or "concerning" should be avoided. The *abstract* should be at least one complete sentence, and at most 300 words. Included with the footnotes to the paper should be the 2010 *Mathematics Subject Classification* representing the primary and secondary subjects of the article. The classifications are accessible from www.ams.org/msc/. The Mathematics Subject Classification footnote may be followed by a list of *key words and phrases* describing the subject matter of the article and taken from it. Journal abbreviations used in bibliographies are listed in the latest *Mathematical Reviews* annual index. The series abbreviations are also accessible from www.ams.org/msnhtml/serials.pdf. To help in preparing and verifying references, the AMS offers MR Lookup, a Reference Tool for Linking, at www.ams.org/mrlookup/.

Electronically prepared manuscripts. The AMS encourages electronically prepared manuscripts, with a strong preference for $\mathcal{AMS}$-LaTeX. To this end, the Society has prepared $\mathcal{AMS}$-LaTeX author packages for each AMS publication. Author packages include instructions for preparing electronic manuscripts, samples, and a style file that generates the particular design specifications of that publication series. Though $\mathcal{AMS}$-LaTeX is the highly preferred format of TeX, author packages are also available in $\mathcal{AMS}$-TeX.

Authors may retrieve an author package for *Memoirs of the AMS* from www.ams.org/journals/memo/memoauthorpac.html or via FTP to ftp.ams.org (login as anonymous, enter your complete email address as password, and type cd pub/author-info). The

AMS Author Handbook and the *Instruction Manual* are available in PDF format from the author package link. The author package can also be obtained free of charge by sending email to tech-support@ams.org or from the Publication Division, American Mathematical Society, 201 Charles St., Providence, RI 02904-2294, USA. When requesting an author package, please specify $\mathcal{AMS}$-LaTeX or $\mathcal{AMS}$-TeX and the publication in which your paper will appear. Please be sure to include your complete mailing address.

After acceptance. The source files for the final version of the electronic manuscript should be sent to the Providence office immediately after the paper has been accepted for publication. The author should also submit a PDF of the final version of the paper to the editor, who will forward a copy to the Providence office.

Accepted electronically prepared files can be submitted via the web at www.ams.org/submit-book-journal/, sent via FTP, or sent on CD to the Electronic Prepress Department, American Mathematical Society, 201 Charles Street, Providence, RI 02904-2294 USA. TeX source files and graphic files can be transferred over the Internet by FTP to the Internet node ftp.ams.org (130.44.1.100). When sending a manuscript electronically via CD, please be sure to include a message indicating that the paper is for the *Memoirs*.

Electronic graphics. Comprehensive instructions on preparing graphics are available at www.ams.org/authors/journals.html. A few of the major requirements are given here.

Submit files for graphics as EPS (Encapsulated PostScript) files. This includes graphics originated via a graphics application as well as scanned photographs or other computer-generated images. If this is not possible, TIFF files are acceptable as long as they can be opened in Adobe Photoshop or Illustrator.

Authors using graphics packages for the creation of electronic art should also avoid the use of any lines thinner than 0.5 points in width. Many graphics packages allow the user to specify a "hairline" for a very thin line. Hairlines often look acceptable when proofed on a typical laser printer. However, when produced on a high-resolution laser imagesetter, hairlines become nearly invisible and will be lost entirely in the final printing process.

Screens should be set to values between 15% and 85%. Screens which fall outside of this range are too light or too dark to print correctly. Variations of screens within a graphic should be no less than 10%.

Inquiries. Any inquiries concerning a paper that has been accepted for publication should be sent to memo-query@ams.org or directly to the Electronic Prepress Department, American Mathematical Society, 201 Charles St., Providence, RI 02904-2294 USA.

Editors

This journal is designed particularly for long research papers, normally at least 80 pages in length, and groups of cognate papers in pure and applied mathematics. Papers intended for publication in the *Memoirs* should be addressed to one of the following editors. The AMS uses Centralized Manuscript Processing for initial submissions to AMS journals. Authors should follow instructions listed on the Initial Submission page found at www.ams.org/memo/memosubmit.html.

Algebra, to ALEXANDER KLESHCHEV, Department of Mathematics, University of Oregon, Eugene, OR 97403-1222; e-mail: klesh@uoregon.edu

Algebraic geometry, to DAN ABRAMOVICH, Department of Mathematics, Brown University, Box 1917, Providence, RI 02912; e-mail: amsedit@math.brown.edu

Algebraic geometry and its applications, to MINA TEICHER, Emmy Noether Research Institute for Mathematics, Bar-Ilan University, Ramat-Gan 52900, Israel; e-mail: teicher@macs.biu.ac.il

Algebraic topology, to ALEJANDRO ADEM, Department of Mathematics, University of British Columbia, Room 121, 1984 Mathematics Road, Vancouver, British Columbia, Canada V6T 1Z2; e-mail: adem@math.ubc.ca

Combinatorics, to JOHN R. STEMBRIDGE, Department of Mathematics, University of Michigan, Ann Arbor, Michigan 48109-1109; e-mail: JRS@umich.edu

Commutative and homological algebra, to LUCHEZAR L. AVRAMOV, Department of Mathematics, University of Nebraska, Lincoln, NE 68588-0130; e-mail: avramov@math.unl.edu

Complex analysis and harmonic analysis, to MALABIKA PRAMANIK, Department of Mathematics, 1984 Mathematics Road, University of British Columbia, Vancouver, BC, Canada V6T 1Z2; e-mail: malabika@math.ubc.ca

Differential geometry and global analysis, to CHRIS WOODWARD, Department of Mathematics, Rutgers University, 110 Frelinghuysen Road, Piscataway, NJ 08854; e-mail: ctw@math.rutgers.edu

Dynamical systems and ergodic theory and complex analysis, to YUNPING JIANG, Department of Mathematics, CUNY Queens College and Graduate Center, 65-30 Kissena Blvd., Flushing, NY 11367; e-mail: Yunping.Jiang@qc.cuny.edu

Functional analysis and operator algebras, to NATHANIEL BROWN, Department of Mathematics, 320 McAllister Building, Penn State University, University Park, PA 16802; e-mail: nbrown@math.psu.edu

Geometric analysis, to WILLIAM P. MINICOZZI II, Department of Mathematics, Johns Hopkins University, 3400 N. Charles St., Baltimore, MD 21218; e-mail: trans@math.jhu.edu

Geometric topology, to MARK FEIGHN, Math Department, Rutgers University, Newark, NJ 07102; e-mail: feighn@andromeda.rutgers.edu

Harmonic analysis, representation theory, and Lie theory, to E. P. VAN DEN BAN, Department of Mathematics, Utrecht University, P.O. Box 80 010, 3508 TA Utrecht, The Netherlands; e-mail: E.P.vandenBan@uu.nl

Logic, to STEFFEN LEMPP, Department of Mathematics, University of Wisconsin, 480 Lincoln Drive, Madison, Wisconsin 53706-1388; e-mail: lempp@math.wisc.edu

Number theory, to JONATHAN ROGAWSKI, Department of Mathematics, University of California, Los Angeles, CA 90095; e-mail: jonr@math.ucla.edu

Number theory, to SHANKAR SEN, Department of Mathematics, 505 Malott Hall, Cornell University, Ithaca, NY 14853; e-mail: ss70@cornell.edu

Partial differential equations, to GUSTAVO PONCE, Department of Mathematics, South Hall, Room 6607, University of California, Santa Barbara, CA 93106; e-mail: ponce@math.ucsb.edu

Partial differential equations and dynamical systems, to PETER POLACIK, School of Mathematics, University of Minnesota, Minneapolis, MN 55455; e-mail: polacik@math.umn.edu

Probability and statistics, to RICHARD BASS, Department of Mathematics, University of Connecticut, Storrs, CT 06269-3009; e-mail: bass@math.uconn.edu

Real analysis and partial differential equations, to WILHELM SCHLAG, Department of Mathematics, The University of Chicago, 5734 South University Avenue, Chicago, IL 60615; e-mail: schlag@math.uchicago.edu

All other communications to the editors, should be addressed to the Managing Editor, ROBERT GURALNICK, Department of Mathematics, University of Southern California, Los Angeles, CA 90089-1113; e-mail: guralnic@math.usc.edu.

Titles in This Series

999 **Montserrat Casals-Ruiz and Ilya Kazachkov,** On systems of equations over free partially commutative groups, 2011

998 **Guillaume Duval,** Valuations and differential Galois groups, 2011

997 **Hideki Kosaki,** Positive definiteness of functions with applications to operator norm inequalities, 2011

996 **Leonid Positselski,** Two kinds of derived categories, Koszul duality, and comodule-contramodule correspondence, 2011

995 **Karen Yeats,** Rearranging Dyson-Schwinger equations, 2011

994 **David Bourqui,** Fonction zêta des hauteurs des variétés toriques non déployées, 2011

993 **Wilfrid Gangbo, Hwa Kil Kim, and Tommaso Pacini,** Differential forms on Wasserstein space and infinite-dimensional Hamiltonian systems, 2011

992 **Ralph Greenberg,** Iwasawa theory, projective modules, and modular representations, 2011

991 **Camillo De Lellis and Emanuele Nunzio Spadaro,** Q-valued functions revisited, 2011

990 **Martin C. Olsson,** Towards non-abelian p-adic Hodge theory in the good reduction case, 2011

989 **Simon N. Chandler-Wilde and Marko Lindner,** Limit operators, collective compactness, and the spectral theory of infinite matrices, 2011

988 **R. Lawther and D. M. Testerman,** Centres of centralizers of unipotent elements in simple algebraic groups, 2011

987 **Mike Prest,** Definable additive categories: Purity and model theory, 2011

986 **Michael Aschbacher,** The generalized fitting subsystem of a fusion system, 2011

985 **Daniel Allcock, James A. Carlson, and Domingo Toledo,** The moduli space of cubic threefolds as a ball quotient, 2011

984 **Kang-Tae Kim, Norman Levenberg, and Hiroshi Yamaguchi,** Robin functions for complex manifolds and applications, 2011

983 **Mark Walsh,** Metrics of positive scalar curvature and generalised Morse functions, part I, 2011

982 **Kenneth R. Davidson and Elias G. Katsoulis,** Operator algebras for multivariable dynamics, 2011

981 **Dillon Mayhew, Gordon Royle, and Geoff Whittle,** The internally 4-connected binary matroids with no $M(K_{3,3})$-Minor, 2010

980 **Liviu I. Nicolaescu,** Tame flows, 2010

979 **Jan J. Dijkstra and Jan van Mill,** Erdős space and homeomorphism groups of manifolds, 2010

978 **Gilles Pisier,** Complex interpolation between Hilbert, Banach and operator spaces, 2010

977 **Thomas Lam, Luc Lapointe, Jennifer Morse, and Mark Shimozono,** Affine insertion and Pieri rules for the affine Grassmannian, 2010

976 **Alfonso Castro and Víctor Padrón,** Classification of radial solutions arising in the study of thermal structures with thermal equilibrium or no flux at the boundary, 2010

975 **Javier Ribón,** Topological classification of families of diffeomorphisms without small divisors, 2010

974 **Pascal Lefèvre, Daniel Li, Hervé Queffélec, and Luis Rodríguez-Piazza,** Composition operators on Hardy-Orlicz space, 2010

973 **Peter O'Sullivan,** The generalised Jacobson-Morosov theorem, 2010

972 **Patrick Iglesias-Zemmour,** The moment maps in diffeology, 2010

971 **Mark D. Hamilton,** Locally toric manifolds and singular Bohr-Sommerfeld leaves, 2010

970 **Klaus Thomsen,** C^*-algebras of homoclinic and heteroclinic structure in expansive dynamics, 2010

969 **Makoto Sakai,** Small modifications of quadrature domains, 2010

TITLES IN THIS SERIES

968 **L. Nguyen Van Thé,** Structural Ramsey theory of metric spaces and topological dynamics of isometry groups, 2010

967 **Zeng Lian and Kening Lu,** Lyapunov exponents and invariant manifolds for random dynamical systems in a Banach space, 2010

966 **H. G. Dales, A. T.-M. Lau, and D. Strauss,** Banach algebras on semigroups and on their compactifications, 2010

965 **Michael Lacey and Xiaochun Li,** On a conjecture of E. M. Stein on the Hilbert transform on vector fields, 2010

964 **Gelu Popescu,** Operator theory on noncommutative domains, 2010

963 **Huaxin Lin,** Approximate homotopy of homomorphisms from $C(X)$ into a simple C^*-algebra, 2010

962 **Adam Coffman,** Unfolding CR singularities, 2010

961 **Marco Bramanti, Luca Brandolini, Ermanno Lanconelli, and Francesco Uguzzoni,** Non-divergence equations structured on Hörmander vector fields: Heat kernels and Harnack inequalities, 2010

960 **Olivier Alvarez and Martino Bardi,** Ergodicity, stabilization, and singular perturbations for Bellman-Isaacs equations, 2010

959 **Alvaro Pelayo,** Symplectic actions of 2-tori on 4-manifolds, 2010

958 **Mark Behrens and Tyler Lawson,** Topological automorphic forms, 2010

957 **Ping-Shun Chan,** Invariant representations of GSp(2) under tensor product with a quadratic character, 2010

956 **Richard Montgomery and Michail Zhitomirskii,** Points and curves in the Monster tower, 2010

955 **Martin R. Bridson and Daniel Groves,** The quadratic isoperimetric inequality for mapping tori of free group automorphisms, 2010

954 **Volker Mayer and Mariusz Urbański,** Thermodynamical formalism and multifractal analysis for meromorphic functions of finite order, 2010

953 **Marius Junge and Javier Parcet,** Mixed-norm inequalities and operator space L_p embedding theory, 2010

952 **Martin W. Liebeck, Cheryl E. Praeger, and Jan Saxl,** Regular subgroups of primitive permutation groups, 2010

951 **Pierre Magal and Shigui Ruan,** Center manifolds for semilinear equations with non-dense domain and applications to Hopf bifurcation in age structured models, 2009

950 **Cédric Villani,** Hypocoercivity, 2009

949 **Drew Armstrong,** Generalized noncrossing partitions and combinatorics of Coxeter groups, 2009

948 **Nan-Kuo Ho and Chiu-Chu Melissa Liu,** Yang-Mills connections on orientable and nonorientable surfaces, 2009

947 **W. Turner,** Rock blocks, 2009

946 **Jay Jorgenson and Serge Lang,** Heat Eisenstein series on $SL_n(C)$, 2009

945 **Tobias H. Jäger,** The creation of strange non-chaotic attractors in non-smooth saddle-node bifurcations, 2009

944 **Yuri Kifer,** Large deviations and adiabatic transitions for dynamical systems and Markov processes in fully coupled averaging, 2009

943 **István Berkes and Michel Weber,** On the convergence of $\sum c_k f(n_k x)$, 2009

942 **Dirk Kussin,** Noncommutative curves of genus zero: Related to finite dimensional algebras, 2009

For a complete list of titles in this series, visit the
AMS Bookstore at **www.ams.org/bookstore/**.